American Post-Conflict Educational Reform

American Post-Conflict Educational Reform

From the Spanish-American War to Iraq

Edited by
Noah W. Sobe

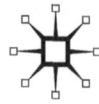

AMERICAN POST-CONFLICT EDUCATIONAL REFORM
Copyright © Noah W. Sobe, 2009.

Softcover reprint of the hardcover 1st edition 2009 978-0-230-61592-2

All rights reserved.

First published in 2009 by
PALGRAVE MACMILLAN®
in the United States—a division of St. Martin's Press LLC,
175 Fifth Avenue, New York, NY 10010.

Where this book is distributed in the UK, Europe and the rest of the world,
this is by Palgrave Macmillan, a division of Macmillan Publishers Limited,
registered in England, company number 785998, of Houndmills,
Basingstoke, Hampshire RG21 6XS.

Palgrave Macmillan is the global academic imprint of the above companies
and has companies and representatives throughout the world.

Palgrave® and Macmillan® are registered trademarks in the United States,
the United Kingdom, Europe and other countries.

ISBN 978-1-349-37951-4   ISBN 978-0-230-10145-6 (eBook)
DOI 10.1057/9780230101456

Library of Congress Cataloging-in-Publication Data

    American post-conflict educational reform : from the Spanish-
American War to Iraq / edited by Noah W. Sobe.
        p. cm.
    Includes bibliographical references and index.
    1. Educational assistance, American. 2. Educational change.
3. Conflict management. I. Sobe, Noah W., 1971–

LB2283.A47 2009
379.1'290973—dc22   2009014130

A catalogue record of the book is available from the British Library.

Design by Newgen Imaging Systems (P) Ltd., Chennai, India.

First edition: December 2009

10 9 8 7 6 5 4 3 2 1

CONTENTS

Part I   Introduction

Part II   The "American Century" Begins

Part III   Promises of Modernity and Abundance

PART I

Introduction

CHAPTER ONE

American Imperatives, Educational Reconstruction and the Post-Conflict Promise

NOAH W. SOBE

In the beauty of the lilies Christ was born across the sea,
With a glory in his bosom that transfigures you and me:
As he died to make men holy, let us die to make men free,
—Julia Ward Howe,
"Battle Hymn of the Republic" (1862)[1]

In one telling of the story, the transatlantic transfer of European civilization has created a new kind of democratic people who are haunted by an obligation to bring a similar conversion to other peoples around the world. After observing a review of Union troops in the midst of the American Civil War, the ardent abolitionist Julia Ward Howe penned the "Battle Hymn of the Republic,"[2] which proposed that the distinctive American mission of war was to "to make men free." Though this notion has not gone unquestioned and though the song itself has been satirized many times—for example, by Mark Twain who in 1901 suggested, "as Christ died to make men holy, let men die to make us rich"[3]—the ideas it expresses have shown a curious resiliency across American history. The notion that Americans have a messianic duty to make others free recurs as a topic of cultural reflection and regularly informs policy and actions. National symbols and national narratives come out in spades in times of war, yet war sometimes also provides moments of clarity and insight into the composition of social imaginaries.[4] Howe's battle hymn provides one crystalline image of the American relationship to military conflict; the "Star-Spangled Banner," which became the U.S. national anthem in 1931, provides another.

When Francis Scott Key, an American lawyer and amateur poet, contemplated the early light on the morning of September 4, 1814, he was aboard a British prisoner-of-war ship and observing the attempted invasion of Baltimore. Key's verse was set by his brother-in-law to the tune of an old drinking song and by the end of that month had been published in newspapers across the United States. As historian Robert A. Ferguson and others have pointed out, this text makes a strange choice for a national anthem. Its first stanzas feature a series of hesitant questions—"can you see? / does that star-spangled banner yet wave?"— hardly a rousing, self-affirming national hymn on the face of it. In Key's version, it was in the light of exploding shells that he was able to see the American flag flying over a besieged Fort McHenry. However, the ultimate outcome of the battle is revealed at the end of the second, rarely sung, stanza when the light of day showed that the American flag was still flying. ("Now it catches the gleam of the morning's first beam, In full glory reflected now shines in the stream"—see figure 1.1). Ferguson argues that this combination of uncertainty and conviction captures in a nutshell the special meaning that the Enlightenment took on in America. The song offers a narrative of origins that symbolically places the constitution of America in a moment of violence. The

> O ! say can you see by the dawn's early light,
> What so proudly we hailed at the twilight's last gleaming,
> Whose broad stripes and bright stars through the perilous fight,
> O'er the ramparts we watch'd, were so gallantly streaming?
> And the Rockets' red glare, the Bombs bursting in air,
> Gave proof through the night that our Flag was still there;
> O ! say does that star-spangled Banner yet wave,
> O'er the Land of the free, and the home of the brave?
>
> On the shore dimly seen through the mists of the deep,
> Where the foe's haughty host in dread silence reposes,
> What is that which the breeze, o'er the towering steep,
> As it fitfully blows, half conceals, half discloses?
> Now it catches the gleam of the morning's first beam,
> In full glory reflected new shines in the stream,
> 'Tis the star spangled banner, O ! long may it wave
> O'er the lard of the free and the home of the brave.

Figure 1.1   Initial two stanzas of the U.S. national anthem as first published. (Published as a broadside under the title "Defence of Fort M'Henry," September 1814. Reprinted with permission of the Baltimore Historical Society.)

anthem also plays on the metaphor of light, which here resolves the uncertain. That knowledge and science could illuminate the unknown was the central Enlightenment metaphor and one particularly suitable for Americans engrossed in settling and expanding across an unfamiliar continent. Ferguson proposes that while the Enlightenment trades in ringing affirmations, "its deepest meanings lie in the uncertain struggle of light against darkness."[5] In the Republican experiment that was the early United States, the struggle between chaos and order took on practical urgency. And, as is well known, public schooling was proposed as one measure which could ensure that reason and reasonableness would be widely distributed across the population, thus helping to guarantee the success of a political order based on self-determination. While there is plenty to suggest that even in the colonial period publicly organized schooling served other social purposes as well, this argument linking the school and American democracy has proven extremely durable. Time and again it has been endorsed with boundless confidence, even as its proponents simultaneously have been haunted by a consuming anxiety about its effectiveness and practical implementation. The U.S. national anthem exhibits a similar anxiety—again, not over what to do or which principles to uphold, but over whether Americans will succeed in their efforts. Accordingly, the moment when the smoke of battle clears and the first rays of the new day illuminate the landscape proves to be a super-charged, revelatory moment. By the dawn's early light all is renewed and all is possible.

I begin this introductory chapter with a discussion of these two popular "national" anthems in order to focus attention on the hope and conviction that in American eyes so commonly accompanies the cessation of military conflict. In the American imaginary, the dawn of a post-conflict era is often construed as a moment of opportunity—an opportunity for emancipation from the past, for wide-scale social reengineering, and for laying the foundations of a stable, peaceful post-conflict order. It is also a pivotal moment for the enactment of a global civilizing mandate.

This volume brings together historians of education and comparative education scholars to examine the ways that the reform of schooling has figured in U.S. post-conflict reconstruction efforts around the globe. While it is not uniquely American to hold that reconstructing education systems is a reliable route to reconstructing societies, educational restructuring has been a core feature of a number of American overseas initiatives, from the Spanish-American War of 1898 through the present. We only examine the reconstruction efforts of a single country in part for the pragmatic reason of presenting a more coherent and complete study than would be possible otherwise. However, our primary reason for focusing on U.S. educational reconstruction projects has to do with America's global prominence in the twentieth century and at the outset of the twenty-first. As discussed at greater length below, this is a period in which American educational initiatives formed one of the means by which a preeminent or, if you prefer, hegemonic position has been established for the United States. The chapters in this volume all focus on "post-conflict settings," though this is approached with flexibility as not all of the sites examined had witnessed the actual destructions of military conflict. Using a variety of approaches, the chapters examine the norms, ideals, strategies, and techniques embedded in American efforts to rebuild school systems, restructure pedagogical practices, and reform curricula overseas.

Empire, American Schooling, and
Post-Conflict Reconstruction

That noted historian, Charles S. Maier could write in 2006 that "until a few years ago, most historians and commentators who wrote about empire angrily rejected any application of the concept to the U.S. as somehow un-American" is good evidence in and of itself of the deep conflicts between the idea of empire and cherished notions of "benevolence" that are so central to the idealized American self-image.[6] Maier's comment ignores revisionist historians such as William Appleman Williams and Walter LaFeber, not to mention W. E. B. Dubois' penetrating criticism from the beginning of the century.[7] Yet, it demonstrates that much is at stake in characterizing America as imperial or not. This is far more than a rarefied academic question. It is one that has been a concern of historical actors themselves, some of whom embrace and some of whom reject an imperial image and role. At the heart of the issue are questions of coercion and consent, imposition and free choice. In the "exceptionalist" paradigm of American intellectual thought, democratic America was to be different from autocratic Europe, and accordingly there were supposed to be differences in the way that Americans went about aiding and enlightening others.[8]

The chapters in this volume position the relationship between coercion and consent as an issue to be analyzed and explored in the context of specific social and historical settings. As I discuss below, the contributors make use of different theoretical literatures to think about power, influence, and, indeed, empire. However, as an initial entry point it is useful to consider Benjamin Justice's argument in chapter 2 that elements of American education can be characterized as "imperial education" whether deployed in the Philippines, in the postbellum South, or for that matter in Massachusetts. By this, Justice directs our attention to the fallacy of national histories predicated on an a priori, categorical distinction between the "internal" and "external" and instead shows the flexibility, mobility, and regular deployment of insider-outsider dynamics across time. Schools are deeply implicated in the exercise of power and like other institutions such as the family and military are deeply caught up in the policing of social and cultural boundaries. They also work on such an intimate level that bodies, attitudes, and actions are brought under tight surveillance, with the result that individuals are "enabled" or "disabled": they are able to access certain social arenas and social goods and simultaneously excluded from others. Along these lines, Justice invites us to consider the importance of the positionality of the educator, specifically what it has meant that the United States so frequently claims to be acting in the guise of teacher. As numerous scholars have also proposed, this speaks to expanding beyond the narrow designation of empire as the formal acquisition of territory to consider the cultures, social categories, and social relations that accompany (and compose) imperial politics.[9] Though we do aim to shed light on questions such as whether an empire is something one "has" or "is," or what the difference is between the "imperial" and the "colonial," these are not questions that this volume seeks to resolve definitively. Our starting point is to look at specific historical instances where one can see these issues being articulated and enacted, challenged and realigned, clarified and blurred.

To fully appreciate the breadth and complexity of American post-conflict educational reconstruction activities from the Spanish-American War to Iraq it is necessary to examine a range of American actors and institutions. The pieces in this collection look at work done by missionaries, government officials, philanthropists, volunteers, military advisors, university professors, and development consultants, among others. We also examine the various—and sometimes competing—institutions that enabled and implemented reconstruction projects. Consequently, the chapters look at both governmental and nongovernmental education projects. In one part, this is in keeping with a recent scholarly thrust in the area of diplomatic history, which holds that American power is projected overseas both through official foreign policy actions and through the activities of voluntary, civil-society organizations.[10] In other part, this is prompted by the idea that these two "spheres" cannot be analyzed in isolation from one another, something that is abundantly evident, for example, in chapter 9, Dana Burde's study of a U.S. Agency for International Development (USAID) funded-project to develop community organizations in Bosnia. Regardless of the importance that liberal political theories grant to civil society as an arena putatively distinct from that of government,[11] the chapters in this volume clearly show how issues of governance and the exercise of power pervade both governmental and nongovernmental reconstruction projects.

The case studies here have been collected together as instances of *American* post-conflict educational reconstruction because they are associated with the United States—associated in the territorial sense of originating from, or relating to, certain select locations in North America—and, associated in the cultural sense of being linked to American social imaginaries. And, as was suggested above in the discussion of the "Battle Hymn of the Republic" and the "Star-Spangled Banner," we find that these imaginaries both inspire/explain reconstruction projects and figure as "fodder" or strategic tools/objects within the projects themselves.

To be sure, there are many ways to study American involvement in overseas education projects. In fact, there is room for and need for considerably more research on the actors and networks, as well as on the ways that school reform projects globally have reflected and advanced American ideals, sensibilities, and cultural/social patterns.[12] In this volume we restrict our analysis to the way this all played out in post-conflict settings. As proposed above, the light of a new day that marks the end of conflict is frequently considered an extremely opportune moment for intervention and deliberate social transformations. Inasmuch as "the dawn's early light" only appears when Americans and their allies are victorious, the post-conflict moment is also commonly attached to the millennial idea that a new era of peace and prosperity has been broached. For much of the twentieth century, cheery optimism enshrouded *Pax Americana* despite any and all lingering anxieties; whether this has been fundamentally reconfigured with the War on Terror remains to be seen.

Not all of the sites examined in this volume had seen actual fighting, but across nearly all we see the idea that a battle had been won and "liberation" achieved. This discursive positioning was evident in Cuba at the beginning of the twentieth century, as we see in chapter 3 in Jason Yaremko's study of the work of American Protestant missionaries setting up schools in a Cuba "freed" from Spain's autocratic (and Catholic) influence. It was also evident in Eastern Europe in the 1990s as we see in chapter 8 in Laura Perry's analysis of the scholarship American academics produced on educational systems in Eastern Europe in the wake of the Cold War. While the U.S.-Soviet conflict involved ample instances of armed conflict waged by proxies, the "Cold" War never spilled out into direct, armed conflict between the two superpowers. We include its aftermath as one of the post-conflict settings examined in this volume because its end has so often been trumpeted as an "ideological" victory. The Cold War example in fact suggests that, generally speaking, we ought to pay close attention to the ways that military conflicts are coded as conflicts over social ideals and cultural models. The implicit logic is that if success is attained in the battle, the other things must be superior as well.

Approaching American Power, 1898 to the Present

The contributors to this volume take a range of positions when it comes to defining and analyzing "power." Across the book we see an interest in sorting out the entanglements of power and knowledge and making sense of the relations, practices, and privileges that schools produce and reproduce. Yet, the chapters do this in ways that can be quite divergent. The editorial criteria and ambitions for the volume have been sketched out above—we have not aimed to present a uniform theoretical orientation across the chapters. Rather, we propose that there is considerable value to be gained by presenting a variety of analytic approaches. This section quickly introduces each chapter, explains the chronology and composition of the collection, and briefly discusses elements of the theoretical approach taken by each contributor.

The first section of the book is titled "The 'American Century' Begins," in allusion to Henry Luce's famous 1941 essay. It aims to demonstrate the importance that overseas education reform played in constructing the "American Century" as American. In his chapter on the origins of American "imperial Education" and educational initiatives in the Philippines, Benjamin Justice argues that we can properly see the strategies employed there neither as an aberration, nor as a radical new departure, but as fundamentally linked to the means and mechanisms of schooling in America over a long duration. Justice discusses schooling in terms of a power-transfer function whereby education reproduces social inequalities (understood here in terms of differential amounts of power possessed) but also has the potential to furnish tools

that can subvert these power relations. Jason M. Yaremko's chapter
also analyzes American educational initiatives following the Spanish-
American War of 1898. He focuses on the Protestant mission schools
that were established in Cuba alongside an American-administered
public school system, and considers the way that the former worked
to advance U.S. hegemony. Yet, even as the concept of hegemony
evokes the Gramscian position that schools are more in the business of
manufacturing consent than nurturing independent, "free" thinkers,[13]
Yaremko understands this not to be totalizing. He details some of the
ways that the Cuban middle class made use of these American protestant mission schools and, in some cases, put their educations to use in
efforts to challenge the status quo. In the third chapter in this section,
Noah W. Sobe approaches the voluntary and philanthropic reconstruction work undertaken by Americans in Europe after World War I with
an interest in the subjectivities and cultural patterns that were normalized through these projects. At issue, Sobe argues, was the legitimacy of
claims to represent "modern civilization" and all that was progressive
and advanced. He identifies World War I as an important turning point
where European validations of American authority for norms-making
enabled American cultural models to take on increasing force as global
"best-practices." While Sobe does not see this exercise of power issuing forth in a coordinated fashion from a stable central point, he does
identify regulative principles that coalesce to generate authority and a
position of preeminence for the United States.

   The second section of the book includes three chapters grouped
under the heading "Promises of Modernity and Abundance," a title
designed to highlight the fact that by mid-century the United States
was widely recognized as an exemplar of modernity and the originator
of a novel form of consumer society. Charles Dorn and Brian Puaca
look at education reconstruction in Germany over the period 1944–49
and U.S. efforts to "reeducate" the German people away from fascism
toward democracy. In a challenge to some of the existing scholarship,
they argue that American efforts did eventually bear fruit, though only
because Germans adopted, adapted, and reformulated the Americans'
democratizing initiatives. Though their chapter principally focuses on
a report produced by the 1946 U.S. Education Mission to Germany, by
remaining ever mindful of the agency of German educators Dorn and
Puaca are able to illuminate the combinations of factors that make for
democratic change in schools. In the next chapter, Kentaro Ohkura and
Masako Shibata examine efforts by U.S. military authorities to restructure the *Shinto* religion and its relationship to the Japanese state. They

draw on the Foucaultian distinction between "sovereign power" and "productive power" and propose that we consider the ways that religion is productive in enabling people to see, feel, think, and act in particular, effective ways. Okhura and Shibata argue that in promulgating a religion-state separation and in reworking the ways that religion factored into civic/moral education in schools, American reconstruction in Japan after World War II has had lasting influence on the ways that Japanese reflect on the self and on others, and, relatedly, on the configuration of nationalism in contemporary Japan. Thomas Koinzer's chapter on German school reform in the 1960s looks at the activities of both Germans and Americans who saw a continuing need for the democratic educational reconstruction. In a reverse of what we have seen in earlier chapters, Koinzer looks less at ideas emanating out of the United States, but instead focuses on West German educators who traveled to the United States on educational study tours. He focuses on German writings on the "American Way of Life," and discusses the ways that cultural reflections on this notion entered into in German educational reform discourses.

The final four chapters of the book are grouped under the heading "After the Cold War, In the Face of Terror" to direct attention to the changing political and social contexts of American post-conflict educational reconstruction in the last two decades. In some way or another, these pieces all touch on the multiple parties and networks of stakeholders that appear to be increasingly figuring in educational reconstruction projects.[14] Laura Perry examines the research that American academics produced on East European education in the 1990s. She investigates the relationship between knowledge and power by borrowing methodologically and conceptually from Edward Said's analysis of the Orientalist scholarship produced in the West in the nineteenth and twentieth centuries. Said argued that the construction of an exoticized, essentialized other privileged the West,[15] Perry finds that American scholars critiqued East European education systems in much the same vein and preserved for the United States a privileged position as the authority on democratic education. In the subsequent chapter Dana Burde also examines a post–Cold War American educational reconstruction project, however this was one that additionally took place in the aftermath of interethnic armed conflict. She looks at a U.S. Agency for International Development funded attempt to bring American-style Parent Teacher Associations to Bosnian schools. The project, which ultimately failed, was predicated on a belief that nurturing social capital and enhancing civil society would lead to educational improvements

and interethnic cooperation. By closely analyzing funding arrangements and the particular civil-society/nongovernmental organization
model that Americans attempted to foster in Bosnia, Burde explains
how mismatched expectations were produced and a democratization
initiative fell short. In the next chapter M. Ayaz Naseem uses the lens
of geostrategic politics to look at the ways that U.S. strategic interests have affected education policy in Pakistan. He looks first at the
period 1980–87 when U.S. interest in Afghanistan brought in substantial support for the military regime in Pakistan. This was also a
period in which *madrassah* education mushroomed and Naseem finds
U.S. policy complicit in enabling this to happen. The chapter also
looks at the period from 2001 to the present when U.S. interest in
Pakistan revived. Attention has now focused on the problem of violent religious extremism and the United States has supported projects to
"de-Islamize" Pakistani education. However, Naseem argues that while
some *madrassah* reform may be warranted, there are deeper roots to the
militarization of Pakistani textbooks and curriculum that also need to
be addressed. In the final chapter in the volume Kenneth J. Saltman
examines the American involvement in the reconstruction of Iraqi
education, particularly the involvement of for-profit corporations such
as Creative Associates International, Inc. (CAII) in U.S. government
funded projects. The Althusserian notion of the school as Ideological
State Apparatus frames a part of Saltman's analysis and he argues that
while CAII's no-bid contract work in Iraq in 2003–04 marched under
the banner of "democracy promotion," it represented the spread of a
certain strand of neoliberal capitalism and the same kind of privatization and corporatization of schools that is also now being seen in the
United States. No two chapters could possibly capture the breadth of
U.S. post-conflict education reconstructions projects currently ongoing.
However, these two complete the historical arc of the volume and, like
other chapters, identify some of the complex problems inherent in the
promulgation of democracy and freedom through education reform.

Imaginary, Imperative, and Peace-Making

Any educational reconstruction project has multiple stakeholders with
sometimes conflicting agendas. This volume is predicated on the conviction that the effects of a reform are not adequately analyzed solely in
relation to policymakers' intents but need to be analyzed in relation to
the full range of cultural, social, economic, and political repercussions

that follow from educational interventions. We certainly allow that educational reconstruction in post-conflict settings can be improved through best-practices research on the design, implementation, and evaluation of reforms. However, we also maintain that these technical solutions will only be partial until we have a better historical understanding of the linkages between education reform and post-conflict peace building.

As discussed above, this volume looks at the ways that the export of American ideals (and democracy) via education reform is held to be a key part of "waging peace." And, as noted, the contributions discuss the ways that education reconstruction projects have advanced America's global prominence (in military, economic, and cultural terms), however they also take seriously the designs for peace and socially ameliorative intentions that are embedded in American educational reconstruction efforts. This volume aims to problematize the American social imaginaries that have generated the sets of imperatives that we see played out in post-conflict settings, while recognizing that the ideals of freedom and democracy still possess nobility and hoping that through perilous nights the promises of peace are still there.

Notes

1. Julia Ward Howe, "The Battle Hymn of the Republic," *Atlantic Monthly* 9, no. 52 (1862).
2. For more discussion of the "Battle Hymn of the Republic" and particularly the song "John Brown's Body" on which it was based see Annie J. Randall, "A Censorship of Forgetting: Origins and Origin Myths of 'Battle Hymn of the Republic,'" in *Music, Power, and Politics*, ed. Annie J. Randall (New York: Routledge, 2005).
3. Mark Twain, "Battle Hymn of the Republic [Brought Down to Date]," in *Mark Twain's Weapons of Satire: Anti-Imperialist Writings on the Philippine-American War*, ed. Jim Zwick (Syracuse, NY: Syracuse University Press, 1901/1992).
4. I use the term social imaginary to refer to bounded collective conceptual understandings that form subjectivities and inform actions. Benedict Anderson's work is well-known for emphasizing the social and material consequence of the imagination; Charles Taylor has recently discussed the significance of the social imaginary in modernity, defining it as "that common understanding that makes possible common practices and a widely shared sense of legitimacy." See Benedict Anderson, *Imagined Communities: Reflections on the Origin and Spread of Nationalism*, revised ed. (London: Verso, 1991); Charles Taylor, *Modern Social Imaginaries* (Durham: Duke University Press, 2004), 23.
5. Robert A. Ferguson, *The American Enlightenment, 1750–1820*, (Cambridge, MA: Harvard University Press, 1997), 25.
6. Charles S. Maier, *Among Empires: American Ascendancy and Its Predecessors* (Cambridge, MA: Harvard University Press, 2006), 2.
7. Walter LaFeber, *The New Empire: An Interpretation of American Expansion, 1860–1898* (Ithaca, NY: Published for the American Historical Association [by] Cornell University Press, 1967); William Appleman Williams, *Empire as a Way of Life: An Essay on the Causes and Character of*

America's Present Predicament, Along with a Few Thoughts about an Alternative (New York: Oxford University Press, 1980). On Dubois see Amy Kaplan, *The Anarchy of Empire in the Making of U.S. Culture, Convergences* (Cambridge, MA: Harvard University Press, 2002), 171–212.

8. Though I am arguing that these democratic considerations took on special importance for Americans, there are plenty of instances of European overseas educational projects where a comparable dynamic was at work: governing and administering an empire through schools was associated with "civilization," the idea being that the more primitive a society, the more force would be relied upon instead of consent. See, inter alia, John Willinsky, *Learning to Divide the World: Education at Empire's End* (Minneapolis, MN: University of Minnesota Press, 1998).

9. See, for example, Amy Kaplan and Donald E. Pease, *Cultures of United States Imperialism, New Americanists* (Durham: Duke University Press, 1993); Paul A. Kramer, *The Blood of Government: Race, Empire, the United States, & the Philippines* (Chapel Hill: University of North Carolina Press, 2006); Ann Laura Stoler, *Haunted by Empire: Geographies of Intimacy in North American History, American Encounters/Global Interactions* (Durham: Duke University Press, 2006).

10. Volker R. Berghahn, "Philanthropy and Diplomacy in the 'American Century,'" *Diplomatic History* 23, no. 3 (1999); M. A. Heiss, "The Evolution of the Imperial Idea and U.S. National Identity," *Diplomatic History* 26, no. 4 (2002): 511–540.

11. See the careful discussion of this in Nikolas S. Rose, *Powers of Freedom: Reframing Political Thought* (Cambridge, United Kingdom; New York: Cambridge University Press, 1999).

12. Some of this is being addressed in the field of comparative education where scholarship on "educational borrowing and lending" has seen a recent resurgence of interest. See David Phillips and Kimberly Ochs, eds., *Educational Policy Borrowing: Historical Perspectives* (Oxford: Symposium Books, 2004); Gita Steiner-Khamsi, ed., *The Global Politics of Educational Borrowing and Lending* (New York: Teachers College Press, 2004). In the American history of education scholarship this has not been extensively treated, with several notable exceptions, such as Jonathan Zimmerman's comprehensive and synthetic treatment of the overseas work of American teachers, and a recent volume edited by Thomas Popkewitz that deals with the networks and actors that promulgated John Dewey's educational ideas globally. See Thomas S. Popkewitz, *Inventing the Modern Self and John Dewey: Modernities and the Traveling of Pragmatism in Education,* 1st ed. (New York: Palgrave Macmillan, 2005); Jonathan Zimmerman, *Innocents Abroad: American Teachers in the American Century* (Cambridge, MA: Harvard University Press, 2006).

13. For a set of comparative education studies oriented around Gramsci's notion of hegemony see, Thomas Clayton, ed., *Rethinking Hegemony, International Studies in Education and Social Change* (Albert Park, Australia: James Nicholas Publishers, 2006).

14. For further discussion of changing norms of "partnership" in contemporary post-conflict education reconstruction see, Noah W. Sobe, "An Historical Perspective on Coordinating Education Post-Conflict: Biopolitics, Governing at a Distance, and States of Exception," *Current Issues in Comparative Education* 9, no. 2 (2007): 45–54.

15. Edward W. Said, *Orientalism,* 1st ed. (New York: Pantheon Books, 1978).

Bibliography

Anderson, Benedict. *Imagined Communities: Reflections on the Origin and Spread of Nationalism.* Revised ed. London: Verso, 1991.

Berghahn, Volker R. "Philanthropy and Diplomacy in the 'American Century.'" *Diplomatic History* 23, no. 3 (1999): 393–419.

Clayton, Thomas ed. *Rethinking Hegemony, International Studies in Education and Social Change.* Albert Park, Australia: James Nicholas Publishers, 2006.

Ferguson, Robert A. *The American Enlightenment, 1750–1820.* 1st ed. Cambridge, MA: Harvard University Press, 1997.

Heiss, M. A. "The Evolution of the Imperial Idea and U.S. National Identity." *Diplomatic History* 26, no. 4 (2002): 511–540.

Howe, Julia Ward. "The Battle Hymn of the Republic." *The Atlantic Monthly* 9, no. 52 (1862): 10.

Kaplan, Amy. *The Anarchy of Empire in the Making of U.S. Culture, Convergences.* Cambridge, MA: Harvard University Press, 2002.

Kaplan, Amy and Donald E. Pease. *Cultures of United States imperialism, New Americanists.* Durham: Duke University Press, 1993.

Kramer, Paul A. *The Blood of Government : Race, Empire, the United States, & the Philippines.* Chapel Hill: University of North Carolina Press, 2006.

LaFeber, Walter. *The New Empire: An Interpretation of American Expansion, 1860–1898.* Ithaca, NY: Published for the American Historical Association [by] Cornell University Press, 1967.

Maier, Charles S. *Among Empires : American Ascendancy and Its Predecessors.* Cambridge, MA: Harvard University Press, 2006.

Phillips, David and Kimberly Ochs, eds. *Educational Policy Borrowing: Historical Perspectives.* Oxford: Symposium Books, 2004.

Popkewitz, Thomas S. *Inventing the Modern Self and John Dewey: Modernities and the Traveling of Pragmatism in Education.* 1st ed. New York: Palgrave Macmillan, 2005.

Randall, Annie J. "A Censorship of Forgetting: Origins and Origin Myths of 'Battle Hymn of the Republic,'" in *Music, Power, and Politics,* ed. Annie J. Randall, 5–24. New York: Routledge, 2005.

Rose, Nikolas S. *Powers of Freedom: Reframing Political Thought.* Cambridge, United Kingdom; New York: Cambridge University Press, 1999.

Said, Edward W. *Orientalism.* 1st ed. New York: Pantheon Books, 1978.

Sobe, Noah W. "An Historical Perspective on Coordinating Education Post-Conflict: Biopolitics, Governing at a Distance, and States of Exception." *Current Issues in Comparative Education* 9, no. 2 (2007): 45–54.

Steiner-Khamsi, Gita ed. *The Global Politics of Educational Borrowing and Lending.* New York: Teachers College Press, 2004.

Stoler, Ann Laura. *Haunted by Empire: Geographies of Intimacy in North American History, American Encounters/Global Interactions.* Durham: Duke University Press, 2006.

Taylor, Charles. *Modern Social Imaginaries.* Durham: Duke University Press, 2004.

Twain, Mark. "Battle Hymn of the Republic [Brought Down to Date]," in *Mark Twain's Weapons of Satire: Anti-Imperialist Writings on the Philippine-American War,* ed. Jim Zwick, 40–41. Syracuse, NY: Syracuse University Press, 1901/1992.

Williams, William Appleman. *Empire as a Way of Life: An Essay on the Causes and Character of America's Present Predicament, Along with a Few Thoughts About an Alternative.* New York: Oxford University Press, 1980.

Willinsky, John. *Learning to Divide the World: Education at Empire's End.* Minneapolis, MN: University of Minnesota Press, 1998.

Zimmerman, Jonathan. *Innocents Abroad: American Teachers in the American Century.* Cambridge, MA: Harvard University Press, 2006.

PART II

The *"American Century" Begins*

CHAPTER TWO

Education at the End of a Gun: The Origins of American Imperial Education in the Philippines

BENJAMIN JUSTICE

Uncle Sam looms aggressively into the frame, palms extended, making an offer that cannot be refused. A soldier or a teacher? A disparate group of Filipinos stands in the other corner—smaller, lower, proud but indecisive. Their leader, an Orientalized depiction of the Filipino revolutionary leader Emilio Aginaldo, considers the intruders. The hand upon his chin supports an infinite frown. "Take your choice," says Uncle Sam. "I have plenty of both." This image from a November 1901 issue of *Puck* magazine captures beautifully the logic of American imperial education (see figure 2.1). While American forces conducted a bloody program of destruction, torture, and killing in the Philippines, pro-imperial enthusiasts at home championed their efforts in the name of civilization and benevolence. The choice—and responsibility—rested with the Filipinos. It was, in the words of the caption, "Up to them."

The American invasion of the Philippines marks the first major U.S. military campaign of the twentieth century. Beginning with the Spanish-American War (and a virtually inadvertent conquest of Spanish colonial forces in Manila) and ending in gradual stages of American withdrawal and Philippine "independence" over the course of the twentieth century, the American presence in the Philippines highlights the seemingly conflicted nature of American educational policy abroad.

Figure 2.1 Puck Magazine Cartoon "IT'S 'UP TO' THEM" from 1901, The subtitle reads, "UNCLE SAM (to Filipinos.)—You can take your choice;—I have plenty of both!" (Drawn by Joseph Keppler, Jr. (1872–1956), *Puck* 50:1290, November 20, 1901, p. 8, [Author's Private Collection]).

On the one hand, heralds of American imperialism trumpeted their presence in terms of helping the people of the Philippines. Americans brought the bounties of their nation: democracy, economic prosperity, and social progress. The American empire would not be colonial or exploitative, but anticolonial and progressive. On the other hand, the actual U.S. invasion met with strong resistance from Filipinos. The American military campaign, lasting over a decade in some areas, was characterized by exceptional brutality, mortality, and scale. At the center of this Janus-faced effort stood the schoolhouse, which served as the cornerstone of the American effort to transform the Philippines into a modern, democratic nation-state. What logic propelled a country steeped in a political ideology of individual choice, rationalism, and democracy to force public education upon another people at the end of a gun?

As the Keppler cartoon conveys so brilliantly, imperial education is an exercise in power. Ostensibly, imperial schools offer those on the margins of empire the ability to join as full members. Schools can turn outsiders into insiders. Education is the promise of power. But imperial schooling is also an expression of power. In the minds of the powerful, education accords them, the educator with prerogatives of moral and intellectual superiority, even as it creates a subordinate role of the educated. Alongside, behind, or nearby the teacher, there is, somewhere, a soldier. The reverse is also true: alongside, behind, or nearby the soldier there is, somewhere, the teacher. American nationalism, the officially sanctioned story that justified the American empire, demanded that in times of expansion the teacher and soldier stand together.

The following essay attempts to understand how the Keppler cartoon made sense to Americans at the turn of the century. In so doing, it also guns for a broader target: to understand how American education in the Philippines served the interest of the American empire. To that end this essay does not explore a growing scholarly trend that examines how the various people of the Philippines rejected, resisted, or reformed American educational policies—all which they did. On the other hand, this argument does reflect an emerging global view of United States history that does not use the lens of nationalism.[1] While all modern empires educate, the ubiquitous and particularist claims to the role of "educator" by American imperial leaders have been, since Protestant Englishmen first arrived at the start of the seventeenth century, one of the exceptional features of the American empire.

Within that long tradition, the American conquest of the Philippines in the name of benevolence marked new developments in the expansion of the empire, and Americans brought their historical understandings of education to bear on situation. Within the continental United States, European immigrants had built a republic based on white supremacy, the appropriation and use of Native American land, and the exploitation of African labor.[2] Among whites, the empire embodied egalitarian, progressive Enlightenment political ideals that rested on the need for an educated citizenry. The annexation of a subordinate Asian colony in 1898 that was not eligible for future statehood or citizenship, and that was not to be cleared of indigenous people and settled by Europeans, challenged traditional American imperial policy. Nevertheless, Americans in 1901 understood these changes to their empire within the context of their past experiences, and education's role in American domestic life figured centrally in the packaging and consumption of rhetoric supporting expansion into the Philippine Islands.

This essay opens and closes with Keppler's image. In between it moves backward in time to reconstruct the historical meaning of formal education in America—in the imperial, as opposed to national sense: the role of formal education in the expansion of power into new territories and peoples. Those who oversaw and largely cheered the conquest of the Philippines were products of the nineteenth century, not the twentieth. They lived in a culture and society steeped in a particular brand of educational imperialism that dated back to the first encounters between Europeans and indigenous people in North America. Enthusiasts of American imperialism embraced the logic of education at the end of a gun not as a new concept for a new century, but as the logical conclusion of an American imperial education they already knew.

The Imperial Context of American Education

Traditionally, historical approaches to American education view it through the lens of nationalism. Formal education outside of the continental United States does not appear in most synthetic narratives of American education, while studies of "imperial education," with perhaps a few notable exceptions, write the story of education outside the continent as being distinct from that within the "domestic" United States.[3] Moreover, groups that pose problems to the traditional notion that the United States was a coherent nation, (particularly indigenous people and African Americans) appear as add-ons or in separate sections. American education, as a unit of analysis, is almost universally studied as a national enterprise.

Within that tradition, historians tend to focus on one of two questions: what educators claim education does for the educated, and what it actually accomplishes in practice. In the latter case, historians of the last several decades have looked at the ways in which the educated have been agents in this process—resisting or reforming formal education to meet their own needs. Throughout much of the twentieth century, studies of American imperial education—almost always conceived of as separate from American education writ large—interpreted the story of imperial education in particular as a triumph of progress and humanitarianism.[4] More recently, revisionists have rejected that interpretation and applied new lenses of analysis: Marxist theories of economic exploitation, cultural imperialism theory, or analyses of race and racism in imperial policy.[5]

One popular formulation of this dual nature of schools—what they claim to do versus what they actually do in practice—posits that the school is a gatekeeping institution. American schools provide (or deny) social mobility; they form class identity, reinforce or challenge racial and gender hierarchy, sexual mores, or they simply provide children with necessary skills for life and work. Schools create insiders and outsiders to what society considers powerful. In the case of imperial education (defined below), American policy makers market schools to subjected peoples, and market their subjugation to those already inside the empire, based on the promise of schools to provide prosperity for all.[6]

While this formulation, insiders versus outsiders, explains one function of schooling, however, it does not examine what education does for the educator *as* educator. Education is a reciprocal process, of course, one that involves both parties. Through most of American history, imperial expansion came with explicit promises of formal education to allow outsiders to become insiders. Until the late nineteenth century, this education was voluntary; by the time of the American invasion of the Philippines, it became mandatory. Either way, however, the promise of schooling was much more important than its ability to offer actual power sharing to groups of people considered to be "beyond the pale" (an expression born of the British Empire in Ireland). The offer of education constructed a duality of teacher/pupil that reinforced the dominance of those at the center of the empire. What schools actually accomplished in practice was of secondary value—indeed when schools rarely did threaten the imperial boundaries of power, turning outsiders into insiders, American society developed new forms of exclusion. By the time the United States occupied the Philippines, this centuries-old pattern of imperial schooling played a key role in how Americans pictured their place in the world.

American Education as Imperial Education

Since the revolution that first cut colonial ties with Great Britain in 1776, American leaders struggled to develop a sense of nationalism to support the federal government and the economic and social systems that it sustained. But just because leaders *wanted* the United States to be a nation does not mean that it was, nor that it would be any time soon. The millions of conquered and enslaved people who lived within the expanding territory claimed by the government were not counted as full members of society (in its *Dred Scott* decision some 81 years later, the U.S. Supreme Court did not even count African Americans as people deserving rights, but as property). Nor, traditionally, were these people counted by historians who need to fit American history into a nationalist narrative. To call the United States a nation in the nineteenth century is to essentially ignore Africans and indigenous people under American domination, except as outsiders. Frederick Douglas famously explained to William Lloyd Garrison in 1847, "I have no love for America, as such; I have no patriotism. I have no country. What country have I? The Institutions of this Country do not know me—do not recognize me as a man."[7] Including the stories of indigenous people and African Americans in the story of the expansion of the United States makes the American narrative an imperial one. A radically democratic, "American" public school in a small, antebellum Ohio village or large New England textile town, for example, stood on territory appropriated from indigenous people and depended on economic prosperity made possible by the exploitation of African slaves. After the Civil War and Reconstruction, legal and social discrimination throughout the empire protected white privilege. Those ostensibly democratic schools usually excluded both groups—either forcibly and explicitly, or subtly and implicitly through their curricula and segregated districting.

While a "nation" can be defined as an ethnically, culturally, or ideologically homogeneous people, an "empire" is a political or economic unit that binds many nations or dissimilar peoples together coercively. Scholars disagree about what exactly constitutes an empire, but in general the term describes a system by which a powerful group at the center dominates subordinate groups at the outskirts.[8] This domination may be direct, though military occupation, or indirect, through the threat of force and concomitant economic and political influence. It may even be, according to recent scholars, cultural. Many Americans historically rejected the idea that their nation could be an empire, viewing their state as being unified by a common ideology and exceptional in its rejection of centralized government, while simultaneously ignoring African-American and Native American history and embracing the popular logic of Keppler's cartoon in American foreign policy.[9] (Some still do.)

To call America an empire rather than a nation is neither new nor radical, however. It is a tribute to the success of nineteenth-century common schools that Americans of the twenty-first century do not typically conceive of their government or their economic system as being imperial. Indeed, even among historians of America, the reach of nineteenth-century nationalism is just beginning to fade.[10] But the Founding Fathers of the revolutionary generation wrote often of their "country" as an empire, although they hoped that theirs would be an "empire of liberty" as Jefferson put it. Mass education would prevent this new empire from failing where others—particularly Rome—had in the past: it would provide social stability and acceptance of the imperial order by instilling "virtue." The single greatest architect of American nationalism in the nineteenth century, textbook author Noah Webster, wrote hopefully in 1789 that, "Even supposing that a number of republics, kingdoms or empires, should within a century arise and divide this vast territory; still, the subjects of all will speak the same language, and the consequence of this uniformity will be an intimacy of social intercourse hitherto unknown, and a boundless diffusion of knowledge."[11]

At the same time that they hoped to forge a nation out of their empire—the optimistic meaning of e pluribus unum (from many, one)—the Founding Fathers also inherited a territory and imperial system from Great Britain that rested on the repression of two groups of people in particular—Africans and Native Americans. They crafted and ratified a government that protected the continuing subordination and exploitation of both groups. Nor could they foresee the rise of industrial capitalism and globalism that would profoundly change the nature of the empire in the coming century, at once providing unprecedented wealth and opportunity to millions and withholding it from others. Imperial schooling—particularly American imperial schooling—is a matter of perspective.

Defining Insiders and Outsiders

In their most straightforward aspect, schools not only define who is powerful and who is subordinate, but they also transfer power from one generation to the next. Schools thus define *insiders* and *outsiders*, and conformity offers the promise of access to power to the individual while preserving the status quo. Because hegemony in an empire rests on "distance" from the center, formal education helps to define what the "center" is.[12] That schooling is a transaction in power does not mean that it transfers that power equally and fairly however. Even if individuals do conform to the demands of schooling, those at the center of power can still police the boundaries in other ways, by controlling access to employment for example, or using terrorism and ethnic cleansing. Historically, formal education did not significantly subvert existing patterns of hegemony, though for some individuals at some times it did so powerfully. The power-transfer function of formal education cuts to the heart of one of the historical paradoxes of the American empire: education reproduces power relations in society, but education also offers the tools to subvert those relations. Knowledge (in both senses) is power.

Common or public schools defined insiders and outsiders in several ways, depending on the group and the era. By the mid-nineteenth century, the curriculum of American public schools projected a narrative explaining the empire—they used the word "nation," of course—in ringing tones of cultural superiority, political liberty, economic opportunity, and millennial destiny.[13] Regionalism and sectionalism played a role in mediating local meaning within this aggressive nationalism.[14] Insiders were Protestant, white (Anglo-Saxon being the epitome of whiteness), and economically prosperous. Outsiders were non-Protestant, nonwhite, and poor. Toward the end of the nineteenth century, textbooks emphasized a generic Christianity, rather than Protestantism (according insider status on Catholics) but they also increasingly presented "race" as a scientific hierarchy of people according to their relationship to power in the empire, with Northern Europeans at the top, the doomed noble-savage Indian and the ignoble, savage Africans at the bottom.[15]

Among whites, particularly Northern whites, most elites believed that sending all children to school made them better workers, acquiescent citizens, and would allow them to be sorted by their merit—understood as their proficiency at internalizing the habits and content of imperial schooling. As the industrial revolution led to explosive growth in cities and a reordering of the economy, elites looked to schools to inculcate the necessary habits of "punctuality, regularity, industry, and silence," while at the same time, schools became sorters where the growing middle class could acquire social status and gain access to clerical and managerial jobs. Most elite talk—even negative talk—about white "foreigners" in the antebellum era (e.g., Irish and German Catholics), wanted them in schools, not out of them, to be fitted into the imperial order.[16]

Outsiders had a very different relationship to schooling in the ever-expanding American Empire. Before the Civil War, most slave states banned teaching blacks to read and write. Prevented literacy made mass slave rebellions more difficult, but keeping slaves from being educated served another function too. The revolution from the British Empire in 1776 and subsequent formation of the United States of America depended on fundamental belief that men could govern themselves by right of their capacity to reason. To concede the black woman or man's ability to reason made slavery a glaring hypocrisy. As mass schooling developed alongside the expansion of suffrage in the antebellum decades, common schools in the North faced increasing pressure to exclude blacks, a movement that culminated in the Massachusetts Supreme Court's decision in *Roberts vs. City of Boston* (1850) that public schools could segregate according to race (which served as a model for the infamous *Plessy vs. Ferguson* ruling by the U.S. Supreme Court in 1896). Not surprisingly, blacks sought formal education whenever possible.

After the Civil War, the most profound challenge to American society was the integration of the former slave population into American society. Insiders to the empire largely succeeded in terms that did not violate either the triumphant imperial narrative or the dominant social, political, and economic reality in which a caste system replaced overt slavery. Northern victory required schools, which had largely excluded blacks explicitly or implicitly, to develop new methods for preserving the channels of imperial power. Despite some initial successes at equal education, and the Herculean efforts of blacks to educate themselves, Southern public school districts developed very unequal systems of education for blacks and whites.[17] Across the United States, the popular model for black higher education based on the Hampton Institute, for example, offered what historian James Anderson has called "education for servitude."[18] Nevertheless, as blacks did achieve more education, white society developed increasingly rigid policies of racial exclusion. When the U.S. Supreme Court ruled in *Plessy vs. Ferguson* (1896) that separate facilities were constitutional so long as they were equal, it was protecting the social fortifications white Americans had erected to maintain imperial hierarchies that education alone could no longer buttress.

Justifying Empire

Thinking in terms of insiders and outsiders implies that the results are the purpose of schooling. In the case of imperial schooling, however, the results are often secondary to the more immediate problem of justifying the use of force and exploitation in times of expansion. As Plessy shows, on the eve of the invasion of the Philippines, American society had developed the means within educational institutions *and outside of them* to police the boundaries of the empire. At the same time, despite the overwhelming failure of schooling to offer non-Europeans access to power during the first three centuries of their presence in North America, white Americans prided themselves in the educative quality of their nation and the permeability of American society to individual merit. While schools defined insiders and outsiders for those who attended them (or were not allowed to), the *existence* of schools and the opportunity to attend them (real or imagined), served a more important function. It justified the discrepancy between the nationalistic, liberal claims of the American empire and its often brutal reality. During times of imperial expansion, the British and later American empires used the promise of formal education to liberate and include conquered people. But the promise was rarely as serious as the need to justify the conquest in ways that reinforced the legitimacy of the status quo. Three brief examples—the Praying Indians of colonial Massachusetts, the acquisition of the Western territory in the Revolutionary War, and the Civil War—illustrate this point.

Consider the official seal of the Massachusetts Colony (figure 2.2). Designed nearly three centuries before Keppler's Uncle Sam in the Philippines, the "Massadonian Seal" of 1629 drew on the traditions and ideology of an emerging British Empire that strike a familiar chord. An Indian stands wearing naught but leaves, holding a bow and arrow, surrounded by trees, while the words "Come Over and Help Us" flow from his lips.[19] Since Columbus's first voyage West, Europeans beadily viewed the "New World" as a place to trade, raid, and conquer. For the English, who cut their imperial teeth in the conquest of Ireland in the late sixteenth century, North America offered the promise of an overseas empire to rival Catholic Spain, France, and Portugal, as well as other European powers.[20]

First in Ireland, and then in colonial Virginia and Massachusetts, English colonists developed an imperial ideology that justified their seizure of land in secular and religious terms.[21] The conquered people were primitive savages in their original state of nature. Because they did not have legitimate governments or laws, and because they did not make use their land (transhumance, hunting, and gathering did not count), English colonists could take Indian land without consent.[22] Colonial leaders argued that starting colonies constituted a form of social tutelage, whereby indigenous people would benefit from the increasing prosperity and opportunity that civilization would bring them.[23] In his 1609 book, *Nova Britannia*, for example, Robert Johnson explained that "Our intrusion into their possessions shall tend to their great good.... First in Regard of God the Creator, and of Jesus Christ their Redeemer, if they will believe in him: And secondly, in respect to earthly blessings, whereof they now have no comfortable use." Those "earthly blessings" would accrue to indigenous people as they climbed to higher stages of civilization, starting as a peasant or servant class. In the meantime, "savages" were not entitled to claim their land as

Figure 2.2 Massachusetts Bay Seal. (Nathaniel B. Shurtleff, ed. *Records of the Governor and Company of the Massachusetts Bay in New England* [5 vols. Boston: William White, 1853–54, cover]).

property, which opened up North America not just to political domination, but to the exploitation of natural resources, to settlement, and to colonization.[24] Those who resisted this benevolent tutelage should be treated severely.

The English settlers and investors who journeyed to colonial Massachusetts in the early seventeenth century did not view the Massadonian Seal with irony, but they certainly overlooked the forms of government and conceptions of property that indigenous North Americans did have. And in reality, the Commonwealth of Massachusetts made very little effort to educate Indians in civilization or Christianity.[25] In the most elaborate case of offering education, John Eliot, the minister of Roxbury, went to great lengths to Anglicize and Christianize the Massachusetts Indians. Eliot learned to speak Algonquin, translated religious tracts—including the Bible—and created a series of "praying towns" for Native Americans in Massachusetts, where Indians settled and adopted some modes of "civilization" under the supervision of Europeans (including churches, schools, fixed homes, and European attire), probably in return for what they perceived as personal and cultural preservation in the face of epidemics, war, and expanding white settlement.[26]

Whatever their outer trappings of civilization, however, these red Puritans were still Indians in the eyes of most white ones. Eliot had pushed "Come Over and Help Us" to its logical conclusion, but failed to include the majority in his plans to educate the minority. Englishmen attacked and harassed the Praying Indians throughout the experiment. A large part of the power of English settlers derived from their ability to appropriate Indian lands; to legitimize Indians as full imperial citizens was to strip settlers of their power. This the English laity could not and did not accept, whatever church and civil leaders might say, and however much education might transform Indians into "insiders." Eventually, the colonial government gave in to the pressure during a major Indian war in 1675, and interned some 500 praying Indians to an island in Boston Harbor. Many died of exposure and starvation and their praying towns collapsed.[27] Eliot had embarrassed the Commonwealth of Massachusetts by forcing the issue of Native American tutelage, calling on formal education to turn outsiders into insiders. But popular power rejected the attempt. The transaction between educator and educated had served to legitimize the legal and religious basis of the state, but without ultimately threatening the interest of the English people within it.

A century later, as the British Empire in North America changed hands in the American Revolution, the Continental Congress turned to formal education as a key component of American imperial policy. And again, the actual implementation and effect of imperial education was less important than its existence as a stated aim. Since the end of the

French and Indian War, the British Empire had blocked white settlement into Indian territory west of the Appalachians, choosing to treat Indian societies as legitimate societies, rather than savages in a state of nature. After the Revolution, however, the Continental Congress had no interest in this new approach.[28] On a personal level many revolutionary leaders were themselves deeply involved in western land speculation. On a political level, for a government deeply in debt to its own army (never mind to other governments) the West was everything. Without the revenues of the West, a union of the eastern states had dim prospects for survival.

The land ordinances of the 1780s attempted to regulate white settlement in the Western Territory in order to maximize their economic value and their political viability. As Congress came to focus seriously on the West, squatters became a threat—lawless people who would selfishly snatch up the prime land, while simultaneously repulsing the "right sort" of settler. Eastern writers described them as "rascals," and the "scum and refuse of the continent."[29] As John Jay wrote in a letter to Jefferson in 1786,

> Would it not be wiser gradually to extend our settlements as want of room should make it necessary, than to pitch our tents through the wilderness in a great variety of places, far distant from each other, and from those advantages of education, civilization, law, and government which compact settlements and neighbourhoods afford? Shall we not fill the wilderness with white savages?—and will they not become more formidable to us than the tawny ones which now inhabit it?[30]

And so the Continental Congress got in the education business. To organize the orderly sale of land, they passed the Land Ordinance of 1785. To define the government of the land, and the future states it contained, Congress passed the Northwest Ordinance of 1787. In both documents, Congress used the law to not only shape the landscape and maximize sales but to shape the political and social institutions that settlers created.[31] The Land Ordinances were, in their very design, educational.[32]

And as a part of that educational vision, each document contained provisions for mass education through common schools, organized locally and supported in part by federal land grants.[33] The school provision may be seen as an example of the revolutionary generation's desire for an educated citizenry—one that Jefferson in particular would champion throughout his political and personal life.[34] Looking beyond the text and within the larger utilitarian view (Congress's need to sell land), however, recent historians argue that the provision was intended to attract New Englanders who were used to having public provision for common schools. The Continental Congress wasn't as interested in mass, formal education as it was interested in attracting people who were.[35] It certainly made no effort to follow up on the provisions, and even rejected others.[36]

The provision of public land for public schools did not produce the grand results that Congress extolled. In his analysis of the Northwest Ordinance at the "ground level," historian Carl Kaestle explains that unimproved land was unlikely to raise much rent where such land was abundant and cheap. In some cases, local farmers contracted to use the land for free if they managed to improve it for European-style farming. In other cases, the land just sat, unused, until Congress revised the policy and authorized states to sell school lands in 1826. And in still other cases (the exception, Kaestle argues), the school lands supported the organization of actual, reasonably effective public schools. Until state governments began asserting control and providing leadership and funding in the 1850s and 1860s, local support and operation of schools varied widely according to the wishes and traditions of local majorities, many of which were not inclined to tax themselves to support schools if and when the school fund ran out.[37]

A third example of the imperial function of schooling can be seen in the American Civil War (1861–65), when national government finally trumped state and sectional government, and when, in educational policy, the common school system pioneered in the Northeast and Midwest became the model for the entire United States. That formal education should figure prominently in the Civil War and Reconstruction should not be surprising: by the 1860s governments across Europe were devising or reforming mass educational systems to consolidate national power. Moreover, the common school had become a staple of Northern society and regional identity. What is significant, however, is the way in which formal educational policy worked to absolve the reunited empire of any real responsibility for the incorporation of its largest group of outsiders, freed slaves, into mainstream society on equal terms.

The educational component of Reconstruction was, at times, quite radical. Even before the fighting was over, Northern philanthropists, teachers (white and black), religious congregations, and politicians raised money for, built, and taught schools for freedmen in the South. In the popular imagination Yankee schoolmarms poured into the

South during Reconstruction to rehabilitate Southern whites and civilize Southern blacks. In reality, teachers came for a variety of reasons, and at the heart of the project were African Americans themselves, who placed a high value on gaining education by any means necessary, and made great strides to oversee their own education. Most teachers were local blacks who had educated themselves during or shortly after slavery. With an eye toward more a permanent policy, Congress required Southern states to create systems of mass education for blacks and whites as a condition of readmission to the Union.[38]

Both initially and over time, however, the educational emphasis of Reconstruction failed to remedy racial inequality while at the same time defining the problem in terms of the educator/educated paradigm. Despite their good intentions, most abolitionists and freedmen's aid societies held racist views, and the primary aim of education for freedmen was to preserve the status quo, lest the vast population of freed slaves foment a revolution. More importantly, it placed responsibility for future inequality on blacks, not the social and economic structures of post-slavery America.[39]

African Americans shouldered the burden, making formal education a top postwar priority, both for its inherent value and in recognition of the school's role as a gatekeeping institution. Historian Heather Williams argues that the educational efforts of blacks astonished many white observers and pushed some Southern white elites into supporting black common schools, while black education engendered fear, resentment, and hostility among others. With the collapse of Reconstruction and the rise of a brutal system of racial caste in the American South during the 1880s and 1890s, however, federal and state governments' commitment to the formal education of blacks withered, and with it, much of the promise of equality. Southern states created a two-tiered system of mass education that channeled far more resources to white schools than to black ones, and American society, north and south found other means to police the boundaries between black and white. Offering schools to freedmen had allowed Northern whites the moral luxury of having tried to elevate the African American without offering any serious reform or reparation. In the popular narrative of the American empire, the noble campaign to free the slave had been a triumph of the empire, marked by a failure blacks to keep up their end of the bargain.[40] In that sense, schooling had done its job.

By the end of the nineteenth century and the invasion of the Philippines, most "included" members of the American society would not have explicitly recognized the imperial role of schooling in separating insiders from outsiders, or in legitimizing the empire in times of expansion. Their schools did not teach them this line of analysis, but instead emphasized the glory of Manifest Destiny, the superiority of Anglo Saxon culture, and the importance of subordination to the imperial narrative. Of course formal education did have intrinsic value for all who experienced it, as former slaves knew full well. And schooling for included groups did offer transfer of power, which differed in degree. Education offered blacks more access to American society, albeit within a racial caste system that varied in its rigidity within the regions of the United States. Education for Native Americans was a cruel and nearly complete failure, if judged by the degree to which schools succeeded in providing them access to power and opportunity in mainstream society. On the West Coast, public schools often shut their doors to Chinese and Japanese children, or segregated them with other "undesirables."[41] White Catholics (and smaller non-Protestant groups such as Jews and Mormons) were able to gain the most access to power within the empire among outsiders, either by assimilating into public institutions, including schools, or through a growing array of parochial and private schools and social organizations intended to shelter them from the mainstream and reinforce an alternative to the imperial narrative, but which also gave them political power and social cohesion.[42] The biggest winner was the school itself, which, by the time the United States invaded the Philippines in 1898, stood as an unassailable monument to the empire it served.

The Conquest of the Philippines

The United States did not conquer the Philippines to build schools. On the contrary, as several studies have shown, the McKinley Administration developed its Philippines policy ad hoc, as events in the Spanish-American War (1898) unfolded. War against Spain in the Atlantic theater provided the most immediate motivation to send ships and men to Manila, not to mention longer-term economic interests in expanding American access to the strategic position, raw materials, and markets of Spanish holdings in Latin America and Asia.[43] Despite the efforts of anti-imperialists wary of welcoming tropical territories and nonwhite people into the American fold, the U.S. actions in Asia represented a new phase of imperial policy that had heretofore engaged in either settler colonialism domestically or informal imperialism in Latin America.[44] Once the conquest began in earnest, however, pro-war advocates in the U.S. government and popular press reframed the Philippine issue in familiar terms. The teacher/student dynamic became the literal and metaphorical instrument to justify American expansion. When Joseph Keppler Jr. penned his image of Uncle Sam three years after Dewey's victory in Manila Bay, the conquest of the Philippines had donned the familiar clothes of the American empire: imperialism for education, education in the service of imperialism.

Benevolent Assimilation

President McKinley said publicly and privately that he had no prior interest in the Philippines. Once the conquest of Spain seemed certain, however, the McKinley Administration argued (and the popular press generally agreed) that the American victory over Spain made holding the region a political, even moral necessity.[45] Appealing to his Protestant base, the president claimed that God had spoken to him during prayer, telling him to "uplift, Christianize, and civilize" the Filipino.[46] After a swift victory in the war against Spain, McKinley issued a December 1898 proclamation that described the American policy:

it should be the earnest and paramount aim of the military admin-istration to win the confidence, respect, and affection of the inhabitants of the Philippines by assuring them in every possible way that full measure of individual rights and liberties which is the heritage of a free people, and by assuring them in every possi-ble way that full measure of individual rights and liberties which is the heritage of a free people, and by proving to them that the mission of the United States is one of the *benevolent assimilation*, substituting the mild sway of justice and right for arbitrary rule. In the fulfillment of this high mission, supporting the temperate administration of affairs for the greatest good of the governed, there must be sedulously maintained the strong arm of author-ity, to repress disturbance and to overcome all obstacles to the bestowal of the blessings of good and stable government upon the people of the Philippine Islands under the flag of the United States [emphasis added].[47]

"Benevolent assimilation" rested on popular perception of Filipinos as passive victims of a corrupt and tyrannical Spanish regime; they were a people who were on the one hand primitive and racially inferior, and on the other hand too culturally and ethnically mixed to form a single nation. America would school the Filipino in tapping natural resources and embracing capitalism, in creating good government, and in forming an enlightened nationalism.[48] Of course, some American businesses would benefit handsomely from laying bare the mineral and human wealth of the Islands, and the American military and merchants in the Asian trade would benefit from controlling the strategic value of the archipelago. But the benefit, according to the logic of tutelage, would be mutual.

The largest impediment to the American invasion of the Philippines was not the Spanish, who quickly surrendered, but the Filipinos themselves. Since 1896, a group of Filipino elites from the main island of Luzon had begun a revolution of their own in the name of political and social reform. Their leader, Emilio Aguinaldo, greeted Admiral Dewey and the American forces enthusiastically, assisting in the assault on Spanish forces and trusting American assurances of Filipino independence. American leaders dissembled, however, keeping Filipino revolutionaries separate from the negotiations with the Spanish in Manila, and refusing to recognize Aguinaldo's claims of independence. Fighting soon erupted between Filipinos and Americans, which the McKinley administration labeled an "insurgency" in order to deny that the Filipino resistance was politically legitimate. The task of these insurgents became both military and political, as they attempted to resist the superior American military force while attempting to convince Americans and the world that they were civilized enough to run their own affairs.[49]

Evidence of the inferiority of Filipinos came from many sources in Academe and the American educational establishment. The leading intellectual architect of American understandings of the Philippines was Dean C. Worcester, a University of Michigan zoologist who served on the Philippine Commission. Worcester downplayed the Europeanized and cosmopolitan aspects of the Philippines and focused more on the uncivilized groups, whom he arranged in an elaborate racial hierarchy.[50] In a similar vein, for his annual report for 1898–99 U.S. Education Commissioner Harris printed an essay on the "Intellectual Attainments and Education of the Filipinos," which attempted to synthesize as much Western secondary scholarship as possible. Supposedly scientific and bias-free, the report grouped all people of the Philippines into three categories: "Christianized or civilized peoples," "infidels or heathen," and "Mohammedans" (Muslims). The report conceded that a majority of Filipinos were literate in their own languages, and that there were many more "educated Filipinos" scattered throughout the archipelago than "was generally assumed." But Harris also fell back on stereotyping the intelligence of the people in broad strokes: they were "clever," but were also overly emotional and unable to engage in abstract reasoning. They were musical, mechanically-minded, and superstitious— stereotypes associated with working class immigrants and blacks in the United States.[51]

According to the dominant educational theories of the day, schooling was a process of leading the child through the stages of civilization. By extension, mass education became the means to social evolution, whereby a whole race or nation could evolve. The newly acquired empire in the Caribbean and Southeast Asia provided a special opportunity to prove the merits of American education. In his Annual Report for 1897–98, an enthusiastic William T. Harris (U.S. Commissioner of Education 1889–1906) expounded on the great virtues that formal education could bring to the people of Cuba, Porto Rico [sic], and the Philippines. "It has been said that the child of an American citizen, in a favorable locality, between the years of 1 and 20 passes through all the stages of culture between savage and the highest civilization," he explained. "However this may be, the school in the course of eight years of elementary studies and four years of secondary or higher study fits the youth for understanding and using the instruments of civilization...." Most societies might educate to some degree, he conceded, but the "highest ideal of a civilization is that of a civilization which is engaged constantly in elevating lower classes of people into participation in all that is good and reasonable, and perpetually increasing at the same time their self-activity." In short, an aggressive imperial education policy would, in itself, mark the United States as the most evolved nation on Earth.

> If we can not come into contact with lower civilizations without bringing extermination to their people, we are still far from the goal. It must be our great object to improve our institutions until we can bring blessings to lower peoples and set them on a road to rapid progress. We must take in hand their education.[52]

Harris's theory of cultural recapitulation through education (borrowed heavily from Herbert Spencer) rested on many fundamentally flawed assumptions, not the least of which was that American education did actually offer "blessings," and "rapid progress" to channels of power within the American empire. Among insiders it had done so in limited ways. For the nonwhite peoples of the Philippines, like

Native Americans and African Americans before them, this was a dubious proposition. Reflecting the strength of Anglo-American racism, William Howard Taft, head of the Philippines Commission and first governor of the Philippines wrote candidly that it would take "the training of 50 or a hundred years before [Filipinos] shall ever realize what Anglo-Saxon liberty is."[53]

The Soldier

The dynamic of the teacher/pupil relationship enabled the American Military establishment to ignore the obvious contradiction between impoverishing, relocating, torturing, and killing vast numbers men, women, and children and helping them to be more civilized. Secretary of War Elihu Root explained in an 1899 speech that the American soldier, "brings the schoolbook, the plow, and the Bible. While he leads the forlorn hope of war, he is the advanced guard of liberty and justice, of law and order, and peace and happiness."[54] The popular press generally capitulated in this logic, locating any failure in the American mission on individual Filipino malcontents, rather than on a failure of American policy. The *New York Times,* for example, wrote scathingly of Filipino resistance to the American invasion in February 1899, "We meet these people now, not as pupils at school. But as armed rebels in the field."[55] A pupil's resistance to the teacher's authority must, of course, be punished severely.

Initially the army tried a policy of "friendly," "civilized" war to demonstrate American good intentions.[56] But homegrown American racism flared as Filipino resistance increased. In his study of letters home, Stuart Miller found American soldiers' views of their experience steeped in racial bigotry and enthusiasm for violence, and not the spread of humanitarianism and education. Soldiers commonly referred to people in the Philippines as "niggers," until they developed the term "gugu" (which later generations of GIs would transform into "gook" to describe the people of Vietnam.) The term *nigger* as a default label for racial outsiders was not the only problem. Miller explains. Nearly all high ranking officers had been Indian fighters in the American West, and had participated in the race war of Native American removal in the name of Manifest Destiny.[57]

Because of the political popularity of the war, and the physical distance of the Philippines from the United States, reports of American brutality and atrocities were slow in coming to the attention of Congress and the popular press. Extreme censorship and other antidemocratic measures also reduced the accountability of military commanders to normal, "civilized" rules of engagement. Indeed, the McKinley administration's policy of not recognizing the Filipino people as being civilized, or their revolution as being legitimate, encouraged commanders to dehumanize their foes. Stuart Miller has found widespread reports of mutilation, torture, civilian murder (including women and children), and taking body parts as trophies. The casualty rate in the Philippine war provides further evidence of its brutality: 15 Filipinos killed for every one wounded. The rate in the American Civil War (and the historical norm) was approximately five to one. American commanders bitterly resented Filipino resistance toward their benevolence, especially when Filipinos employed guerilla tactics, and responded with mass atrocities, including the relocation of entire villages into concentration camps and committing mass killings. The year 1901 saw the worst of the atrocities, though news of these did not immediately reach home; when it did, the Congress and the public had little appetite for it.[58]

The Teacher

The American military presence in the Philippines depended on the institution of political, economic, and educational reforms. In his 1899 annual message to Congress, President McKinley promised that

> no effort will be spared to build up the vast places desolated by war and by long years of misgovernment. We shall not wait for the end of strife to begin the beneficent work. We shall continue, as we have begun, to open the schools and the churches, to set the courts in operation, to foster industry and trade and agriculture, and in every way in our power to make these people whom Providence has brought within our jurisdiction feel that it is their liberty and not our power, their welfare and not our gain, we are seeking to enhance.[59]

The army set up schools in Manila and other areas under American control as military operations allowed. Outside areas of control— particularly in non-Christian areas—the military waged its race war with increasing brutality. American educational policy in the Philippines became analogous to the traditional Indian policy within North America: people must be "civilized" before they could be Christianized (and thus accepted into the imperial fold). Civilization meant adopting Western style economic practices, but also social organization, customs, and above all, submission to imperial authority. Thus American forces engaged in two "educations" in the Philippines: the construction of modern schools and expansion of education for those already conversant in the Spanish imperial package, and a battle of dominance/submission in "uncivilized" areas. "One of the first and best results of the substitution of American authority for Spanish rule," explained a special essay in the *Annual Report of the U.S. Commissioner of Education*, "is that everywhere throughout the archipelago the schoolhouse follows the flag." No matter that everywhere the army went they were surprised to find schools already in existence. It was the "broad and liberal methods of American education" that would civilize the Filipino.[60]

In reality, the Spanish colonial administration had already built a system of imperial schools throughout the archipelago, the result of late nineteenth century educational reforms in the Spanish Empire, and efforts all over the Western world and its empires. In higher education, the Philippines contained a university, several regional colleges, and handful of private academies and professional schools (colleges enrolled 8,000 students as of 1896). Clergy also operated 67 Latin grammar schools, most on the island of Luzon, as well as normal schools and academies. Just as in the United States, race hierarchy played a key role within the Spanish regime in the Philippines, with people of Spanish and "mixed-blood" enjoying the privileges of power, and access to higher education. In terms of mass education, Spanish law required single-sex primary schools for boys and girls in all towns with a population of at least 5,000 inhabitants, and three in towns of 10,000. The curriculum included the three "R"s (reading, 'riting, and 'rithmatic), plus Spanish, Catholic Church doctrine, and agriculture.[61]

In practice, however, the system rarely met the legal standard, and functioned to maintain a large, poorly educated class of peasants and a small, well-educated elite class of mostly mestizo Filipinos, who went to finish their education abroad. Regional governments operated roughly half of the required schools—a total of approximately 2,200 in the year 1896, with an enrollment of less than 200,000 (of estimated total Philippine population of seven million). In many areas, Spanish was taught badly or even prohibited. The law made schooling compulsory but local officials did not enforce it and school attendance was small. The pedagogy was very traditional, requiring rote recitation and absolute obedience, and focused strongly on religion. "Native" men developed a culture of resistance to schooling, and viewed young men who went to school too much, or who went abroad, as sellouts and targets for violence.[62]

The temporary schools conducted by the army during the initial phase of the American invasion were no better (and probably worse) and served a more symbolic role than a real one. Despite the usual disruptions brought on by war, General Otis, a Civil War hero, took pains to set up schools in secure areas, opening as many as a thousand by 1900. Army chaplains typically served as superintendents, and soldiers as teachers when civilians could not be found. Americans usually conducted these schools in existing schoolhouses, but struggled to transplant secular, progressive American schooling into the Philippine context. In a 1900 report, the American Philippine Commission described these schools as "poor," "aimless," and "chaotic," lacking teachers, materials, appropriate books, and a clear curriculum or purpose, except to show evidence of American goodwill.[63] They were, quite literally, schools for the sake of schooling, as the head of the military's Department of Public Instruction himself conceded.[64]

In the summer of 1900 the American Philippine Commission hired a full-time, permanent head of education for the Philippines, and by January of 1901established a formal, civilian Bureau of Education. The bureau's regulations and policies mimicked many aspects of the American educational system, including organizing local boards of trustees, employing American textbooks, American pedagogical methods, banishing explicit religious instruction during regular school hours, and teaching English, to name a few. But true to the paternalist ideology of the American occupation, the system concentrated power at the top, with Americans, who were at liberty to appoint, override, and dismiss local Filipino school officials.[65]

The boldest move included the commitment of a force of 1,000 American teachers, recruited at competitive salaries to operate schools and train Filipino teachers. Known as *Thomasites* after the converted military transport many took in their journey from the United States, these opportunistic and optimistic teachers resembled the Peace Corps volunteers of later generations. Their motivations ranged from a genuine desire to help Filipinos to the need for a job or the love of adventure. They came from all over the United States, many from elite colleges and universities including the universities of Michigan, California, and Chicago, and private universities like Harvard, Cornell, and Yale. Despite a host of problems, 845 arrived in the first year. Of these some got sick, died, got married, or gave up. Their situations were often very difficult, plagued by disease, unfamiliar food, hostile locals, incompetent or uncooperative co-teachers, violence from Filipino revolutionaries, isolation, and pay delays. After the first five years, 42 American teachers had died: causes included dysentery and cholera (15); killed by ladrones (6); drowned (3); and suicide (2).[66] Nevertheless, the Thomasite mission played a powerful role in the American imagination of the triumph of "benevolent assimilation."

Despite the genuine effort of paternalistic American educationists to create a system of mass education in the Philippines, however, their policies lacked coherence, vision, and an understanding of the appropriate needs of the diverse peoples of the archipelago. The first general superintendent of education in the Philippines lasted only two years, in which time he drafted four different curricula that pushed, unsuccessfully, an "industrial education" model closely resembling that of the school district in Massachusetts where he had formerly been a principal. His successor favored education for political liberation and democracy, but that model did not last under pressure from elites and the American colonial administration. In the end, the industrial model "triumphed." Moreover, American rulers in the Philippines were never willing to commit the resources necessary to build an adequate infrastructure to enroll all school-aged children, provide them with enough culturally-relevant materials, or train enough Filipino teachers to teach them.[67]

It's Up To Them

What most Americans knew back home, however, was that the United States had entered the Philippines reluctantly, and had shouldered the White Man's burden in an expression of their cultural superiority. By fall of 1901 war support in the popular press was in full bloom. Joseph Keppler and other cartoonists took up the White Man's pen to convey the optimism, and burden, of the American war in the Philippines. Anti-imperialists rejected the idea of burden, and feared the corruption of American racial and cultural purity, competition from Filipino workers, and the withering effects of imperialism on the free political institutions at home. But they usually accepted the notion of Anglo-Saxon superiority. In one exception, members of the African-American press declared solidarity with the Philippine resistance and wondered when the American government would make good on its promises of liberty and educational opportunity at home.[68] Among the crowd of voices and images supporting the American occupation, however, Keppler conveyed an exceptionally powerful and clear understanding of the role of education in the American empire at the turn of the twentieth century.

Keppler's cartoon has a narrative that can be grouped into three general themes. The first is where the Western eye, trained to read left-to-right, engages the page: on the upper left. Enter the iconic images of the American empire: Uncle Sam, the soldier, and the schoolteacher. All are white. The soldier and teacher have blonde hair, Uncle Sam has blue eyes. The teacher and soldier together resemble the idealized unit of organization in Victorian society: the nuclear family: father and mother supported by the empire. Gender also channels power: the big-bosomed female schoolteacher is the icon of benevolence and civilization, while the muscular male is the source of authority and violence. They are balanced, and cannot exist without each other. Taken together, the threesome form a holy trinity that encapsulates the narrative of American progress: white supremacy resting on a righteous military might tempered by liberal benevolence.

The eye then moves right, and slightly downward, where a group of brown-skinned Filipinos stand clustered together, contemplating Uncle Sam's intrusion. Their disorganized formation and varied postures contrast with the upright regularity of the whites. The Filipinos are culturally diverse (lacking a national identity), perverse (the woman holding a naked child has no discernable husband), and indecisive. In contrast to the modern tools of the whites—the schoolbooks and the Krag-Jorgensen rifle—the Filipinos bear relics of past empires and orders of civilization according to the typical American classifications (Muslim, civilized Christian, and heathen): the Muslim rests on a truncheon, a bedizened Aguinaldo wears a Spanish sword, and another man holds a bow and wears feathers, evoking Native American iconography (the fourth man, in the background, looks similar to Aguinaldo, and probably represents the Christian majority). Not only are the Filipinos depicted as inferior by being located below the Americans positionally; they are depicted as being culturally inferior and in need of order. If the white soldier and the teacher are the mother and father, then the Filipinos are the children.

Third, the viewer takes in the framing and background of the image. Keppler juxtaposes Uncle Sam on one side with trees on the other: the Americans emerge from Uncle Sam, from civilization and strength, while the Filipinos emerge from the wilderness. Both groups stand, in this formulation, as opposite ends of the spectrum of human development. Caught between the two, in the background, lies the Filipino village—present but unformed and insubstantial, like a subject waiting for a predicate.

After assessing the image, the eye moves to the caption, and the viewer takes in the work as a whole. "IT's 'UP TO' THEM." Reads the title. The caption adds, "UNCLE SAM (to Filipinos.)—You can take your choice;—I have plenty of both!" The words close the circle of the imperial narrative. Uncle Sam can only be there if the choice is legitimate: the superiority of the whites makes it so. The superiority of the whites depends on their ability to *be* the educators. It is a point of pride that Uncle Sam has plenty of both—not just a threat but a boast. The dynamic of educator/educated exonerates the Americans of self-interest or antidemocratic motivations. Force, embodied in the soldier, is the consequence of the choice of the educated. Ultimately, formal education is less about the results of the transaction between the two people than the existence of the transaction itself. Like "come over and help us" nearly three centuries earlier, Keppler's image brings together elements of education and power at the micro level—the teacher and what she represents—and the macro level:—Uncle Sam's eager intrusion into the Philippines. As a consequence of this imperial construction of power, the United States was not ultimately responsible for the success or failure of the education it provided. Good students will succeed. It is up to them.

Conclusion

The widespread, triumphant view of formal education in American nationalism made the patterns of American imperial educational policy in the Philippines exceptional among European colonies in Asia.[69] American policy makers from 1898–1901 identified with the British Empire especially, with which they shared a white-supremacist, imperial ideology of Anglo Saxonism that placed America and Britain at the top of a hierarchy of civilization, and gave both empires the mutual responsibility for shouldering the White Man's burden.[70] In their governance of the Muslim region of Mindanao, for example, which the Americans designated as "semicivilized and barbarous," officials used British Malaya as a model, but by comparison only. The Americans rejected the British method of indirect rule, in which the imperial bureaucracy co-opted local elites and created educational systems targeting them, and opted instead for a form of direct rule, which attempted to bypass local elites and to create mass education similar to the common schools of the United States.[71]

Transnational comparisons show the difference. In a careful study of British Burma, Dutch Indonesia, French Vietnam, and the American Philippines, Vince Boudreau found that while all three European powers created limited systems of schooling, the American system was both more complete (offering more education to more people), more thorough (with more programs through secondary and university levels in the Philippines), and more directly tied to opportunities to work in the large colonial bureaucracy.[72] In a similar vein Glenn May reports that imperial schools in Burma enrolled 3.3 percent of the population in 1900, in Netherlands India, 1 percent in 1907, while the American schools in the Philippines enrolled 7 percent of the population in 1907.[73] The American commitment to building schools for their own sake was genuine and vigorous. By 1910, Americans built more than 4,000 schools, enrolled 355,722 elementary school children and 3,400 high school children. By 1920, enrollment was close to a million total, though still not adequate for the number of school-aged children.[74]

The real effect of these schools in relation to American imperial rhetoric, however, was marginal. By 1913, less than half of Filipino teachers possessed an intermediate education, and the typical Filipino child spent two years enrolled in school. As late as 1925, an American commission found that "in the great body of Filipino schools the present methods of teaching reading are so deficient that children have so little facility in reading English on leaving school that there is little guarantee of a functional control over the language in adult life." Likewise, the extensive focus on industrial education, based on the Hampton/Tuskegee model developed during Reconstruction to train blacks for servitude in the United States, endowed Filipino children with skills that had little or no demand in the economy.[75] The sheer size of the American project did benefit many individuals, of course, particularly existing elites and a small middle class. But American-style education in the Philippines did not open up the opportunity for Filipinos to participate in the American empire as insiders, nor did it hasten the rise of a new democratic, egalitarian, and prosperous Philippines. And back in the United States, white society developed new racial classifications to close ranks against Filipino immigrants seeking opportunity.[76]

These later outcomes might have been predicted in 1901—not by looking forward to uncertain trajectory of future American policy, but by looking back to the ways in which the architects of American policy conformed to the traditional understanding of education within the American empire. In an all-too-familiar pattern, American imperialists offered a public education in the context of a brutal race war in which Americans practiced torture and summary execution, operated concentration camps, purposefully destroyed homes and whole villages, and killed an estimated 13 percent of the entire population of the Philippines.[77] The paltry, if relatively prolific, schools set up by the Americans must have seemed small comfort to people who had endured such a catastrophic occupation. Self-interest (and compulsory education laws) led increasing numbers of families to send their children to these schools, though not with results matching the millennial promises of imperial promoters to bring political stability, economic prosperity, insider status, and a fair share of the fruits of the empire. Nevertheless American efforts at spreading schooling, in comparison to the much-maligned Spanish regime, gave great comfort to American war enthusiasts at home who could turn to the art of Keppler and others to understand the glory of their new Asian empire. Whether or not American intentions were benevolent, (and in many cases they were), it was benevolence of an unfortunate sort that defined power in a cultural construct that justified violence against resistance while at the same time exonerating the powerful from accountability for genuine power sharing or social reform.

Notes

1. Norbert Kilian, "New Wine in Old Skins? American Definitions of Empire and the Emergence of a New Concept," in Theories of Empire 1450–1800, ed. David Armitage (Brookfield, VT: Ashgate Publishing, 1998), 307–324. See also Benjamin Justice, "Beyond Nationalism: The Founding Fathers and Educational Universalism in the Early Republic," in Advancing Democracy Through Education? US Influence Abroad and Domestic Practices, ed. Bradley Levinson and Doyle Stevick (Charlotte: Information Age Publishing, 2008): 1–27.
2. On the appropriateness of the claim of "seizure," and the development of theories of property in early modern European thought, see James Tully, "Aboriginal Property and Western Theory: Recovering a Middle Ground," in *Theories of Empire 1450–1800*, ed. David Armitage (Brookfield, VT: Ashgate Publishing, 1998), 345–372.
3. The classic example is Lawrence Cremin, *American Education series* (New York: Harper and Row, 1970, 1980, 1988). For imperial education, see Martin Carnoy, *Education as Cultural Imperialism* (New York: D. McKay, 1974). While Carnoy's Leninist analysis clouds his discussion of race in the United States, the very act of including it was bold and highly significant at the time.
4. While this type of narrative rarely appears anymore, the most recent incarnations include: Stanley Karnow, in *Our Image: America's Empire in the Philippines* (New York: Random House, 1989). John Morgan Gates, *Schoolbooks and Krags: The United States Army in the Philippines, 1898–1902* (Westport, CT: Greenwood Press, 1973).
5. William J. Pomeroy, *American Neocolonialism: Its Emergence in the Philippines and Asia* (New York: International Publishers, 1970); Carnoy, *Education as Cultural Imperialism*; Stuart Creighton Miller, *Benevolent Assimilation: The American Conquest of the Philippines, 1899–1903* (New Haven, CT: Yale University Press, 1982); Glenn Anthony May, *Social Engineering in the Philippines: The Aims, Execution, and Impact of American Colonial Policy, 1900–1913* (Westport, CT: Greenwood Press, 1980); Paul Kramer, *The Blood of Government: Race, Empire, the United States, and the Philippines* (Chapel Hill: UNC Press, 2006); Julian Go, "Introduction: Global Perspectives on the U.S. Colonial State in the Philippines," in *The American Colonial State in the Philippines*, ed. Julian Go and Anne L. Foster (Durham, NC: Duke University Press, 2003), 1–42.

6. On "Insiders versus Outsiders," see, for example, David Tyack, *Seeking Common Ground: Public Schools in a Diverse Society* (Cambridge, MA: Harvard University Press: 2004).

7. Frederick Douglass, "Country, Conscience, and the Anti-Slavery Cause: An Address Delivered in New York, New York, May 11, 1847." *New York Daily Tribune*, May 13, 1847. John Blassingame, ed. *The Frederick Douglass Papers: Series One—Speeches, Debates, and Interviews vol. II* (New Haven, CT: Yale University Press, 1979), 57.

8. Alexander J. Motyl, *Imperial Ends: The Decay, Collapse, and Revival of Empires* (New York: Columbia University Press, 2001); Antonio Negri and Michael Hardt, *Empire (Cambridge, MA: Harvard University Press, 2000).*

9. See, for example, *Andrew Bacevich, American Empire: The Realities and Consequences of U.S. Diplomacy (Cambridge, MA: Harvard University Press, 2004).*

10. See, for example, Thomas Bender, *A Nation Among Nations: America's Place in World History* (New York: Hill and Wang, 2006).

11. Noah Webster, Dissertations on the English Language: With Notes, Historical and Critical. To Which is Added, by way of Appendix, an Essay on a Reformed Mode of Spelling, with Dr. Franklin's Arguments on that Subject (Boston, 1789), reprinted in Wilson Smith, ed., *Theories of Education in Early America 1655–1819* (New York: Bobbs-Merrill Company, 1973), 269. More generally, see Justice, "Beyond Nationalism."

12. Motyl, *Imperial Ends*; Hardt and Negri, *Empire* (Cambridge, MA: Harvard University Press, 2000).

13. Ruth Miller Elson, *Guardians of Tradition: American Schoolbooks in the Nineteenth Century* (Lincoln, NE: University of Nebraska Press, 1964), 1.

14. David Tyack, "Monuments Between Covers: The Politics of Textbooks," *American Behavioral Scientist* 42 no. 6 (March 1999): 922–932.

15. Elson, *Guardians of Tradition*, 65–100.

16. A. B. A., "Education and Crime," *The School Review* 8, no. 1 (January 1900), 42–45; Michael Katz, *The Irony of Early School Reform* (Cambridge, MA: Harvard University Press, 1968).

17. Heather Williams, *Self Taught: African American Education in Slavery and Freedom* (Chapel Hill, NC: University of North Carolina Press, 2005).

18. James D. Anderson, "Education for Servitude: The Social Purposes of Schooling in the Black South, 1870–1930." PhD. Diss. University of Illinois at Urbana Champaign, 1973.

19. The quote itself, "Come over and help us," came from the New Testament, when "a vision appeared to Paul in the night; There stood a man of Macedonia, and prayed him, saying, Come over into Macedonia, and help us." *Acts* 16:9 (King James Version).

20. William Kellaway, *The New England Company 1649–1776: Missionary Society to the Indians* (New York: Longmans, 1961), 1–4. On the concept of the "Ocean Sea," see Bender, *A Nation Among Nations.*

21. Nicholas P. Canny, "The Ideology of English Colonization: From Ireland to America," in *Theories of Empire 1450–1800*, ed. David Armitage (Brookfield, VT: Ashgate Publishing, 1998), 179–202.

22. Tully, "Aboriginal Property and Western Theory," 345–372.

23. John H. Elliott, "The Seizure of Overseas Territories by the European Powers," in *Theories of Empire 1450–1800*, ed. David Armitage (Brookfield, VT: Ashgate Publishing, 1998), 139–158.

24. In his *Second Treatise on Government* (section 49), John Locke explained, "Thus in the beginning all the world was America"; but not all the world had reached the advanced stage of Europe.

25. James H. Merrell, "'The Customs of Our Countrey': Indians and Colonists in Early America," in *Strangers Within the Realm: Cultural Margins of the First British Empire*, ed. Bernard Bailyn and Philip D. Morgan (Chapel Hill, NC: University of North Carolina Press, 1991), 117–156.

26. Dane Morrison, *A Praying People: Massachusetts Acculturation and the Failure of the Puritan Mission, 1600–1690* (New York: Peter Lang, 1995).

27. Jennifer Monaghan, *Learning to Read and Write in Colonial America* (Amherst, MA: University of Massachusetts Press; Worcester: American Antiquarian Society, 2005), 46–80; Richard W. Cogley, *John Elliot's Mission to the Indians before King Philips War* (Cambridge, MA: Harvard University Press, 1999); Norman Earl Tanie, "Education in John Elliot's Indian Utopias, 1646–1675," *History of Education Quarterly* 10 (1970): 308–323; Neal Salisbury, "Red Puritans: The 'Praying Indians' of Massachusetts Bay and John Eliot," *The William and Mary Quarterly* third series 31 no. 1 (January 1974): 27–54.

28. Tully, "Aboriginal Property and Western Theory"; Colin Calloway, *The Scratch of a Pen: 1763 and the Transformation of North America* (New York: Oxford University Press, 2006), 93.

29. Peter S. Onuf, *Statehood and Union: A History of the Northwest Ordinance* (Bloomington, IN: Indiana University Press, 1987), 34–36; Peter S. Onuf, "Liberty, Development, and Union: Visions of the West in the 1780s," *The William and Mary Quarterly* third series 43, no. 2 (April 1986):179–213, 187.

30. John Jay, Letter to Thomas Jefferson, Office of Foreign Affairs, December 14, 1786, John Jay Papers, Columbia University, www.columbia.edu/cu/libraries/inside/working/jay/temp2/letterpress/jay_life_v1/jay_life_v1_26.html (accessed June 19, 2009).

31. Rowland Berthoff, "A Country Open for Neighborhood," *Indiana Magazine of History* 84, no. 1 (March 1988); Onuf, "Liberty, Development, and Union"; Beverly W. Bond, Jr., "American Civilization Comes to the Old Northwest," *The Mississippi Valley Historical Review* 19, no. 1 (June 1932): 3–29.

32. Lawrence Cremin, *American Education: The National Experience 1783–1876* (New York: Harper and Row, 1980), 9–10; Onuf, *Statehood and Union*, 38; David Tyack, Thomas James, and Aaron Benevot, *Law and the Shaping of Public Education, 1785–1954* (Madison, WI: University of Wisconsin Press, 1987), chapter 1; Ruth H. Bloch argues that this form of education was vague and weak, leaving a vacuum to be filled, in the coming decades, by organized religion. See "Battling Infidelity, Heathenism and Licentiousness: New England Missions on the Post-Revolutionary Frontier, 1792–1805," in *The Northwest Ordinance: Essays on its Formulation, Provisions, and Legacy*, ed. Frederick D. Williams (East Lansing, MI: Michigan State University Press, 1989), 39–60.

33. Journals of the Continental Congress, 1774–1789, Friday, July 13, 1787. American Memory Website. See also, Frederick D. Williams, ed., *The Northwest Ordinance: Essays on its Formulation, Provisions, and Legacy* (East Lansing, MI: Michigan State University Press, 1989), Appendix.

34. Carl Kaestle calls this "traditional" view. Carl F. Kaestle, "Public Education in the Old Northwest: Necessary to Good Government and the Happiness of Mankind," *Indiana Magazine of History* 84, no. 1 (March 1988): 60–74.

35. Kaestle, "Public Education in the Old Northwest"; Tyack, James, and Benevot, "Law and the Shaping of Public Education," chapter 1.

36. Congress turned down the Ohio Company's proposal to aside extra land for universities (perhaps fearing the jealousy of existing institutions or, more simply, because it seemed to be overkill).

37. Kaestle, "Public Education in the Old Northwest," 60–74.

38. Williams, *Self Taught*, 96–125; Robert C. Morris, *Reading, 'Riting, and Reconstruction: The Education of Freedmen in the South, 1861–1870* (Chicago, IL: University of Chicago Press, 1981); Ronald Butchart, *Northern Schools, Southern Blacks, and Reconstruction: Freemen's Education, 1862–1875* (Westport, CT: Greenwood Press, 1980).

39. Butchart, *Northern Schools*, 10–16.

40. Ibid; Williams, *Self-Taught*, 174–200.

41. Pomeroy, *American Neocolonialism*, 161. In October of 1906, the San Francisco Board of Education created segregated schools for Japanese, Chinese, and Korean children, together with "children of vicious habits and children suffering from contagious or infectious diseases."

42. Jay Dolan, *American Catholic Parish: A History from 1850 to the Present* (New York: Paulist Press, 1987); John T. McGreevy, *Catholicism and American Freedom: A History* (New York: Norton, 2003).

43. The finest study of the road to war in the Philippines is Miller, *Benevolent Assimilation*. See also: Gates, *Schoolbooks and Krags*; Pomeroy, *American Neocolonialism*; May, *Social Engineering*.

44. Eric Love, *Race Over Empire* (Chapel Hill, NC: University of North Carolina Press, 2004), 170–177.

45. Pomeroy, *American Neocolonialism*, 71; Miller, *Benevolent Assimilation*, 1–12.

46. Miller, *Benevolent Assimilation*, 24; Schirmer, Daniel B. and Stephen Rosskamm Shalom, eds. *The Philippines Reader: A History of Colonialism, Neocolonialism, Dictatorship and Resistance*. Boston: South End Press, 1987, 22–23.

47. The Statutes at Large of the United States of America from March 1897 to March 1899 and Recent Treaties, Conventions, Executive Proclamations, and The Concurrent Resolutions of the Two Houses of Congress, Volume XXX, published by the U.S. Government Printing Office, 1899.

48. May, *Social Engineering*; Pomeroy, *American Neocolonialism*.

49. Kramer, *Blood of Government*, 116.

50. Carla M. Sinopoli and Lars Fogelman, eds., *Imperial Imaginings: The Dean C. Worcester photographic collection of the Philippines, 1890–1913* (Ann Arbor, MI: University of Michigan, 1998); Kramer, *Blood of Government*, 183–184.

51. Annual Reports of the Department of the Interior. Report of the Commissioner of Education. 1898–99, 1595–1638.

52. 1897–98 Report. vol. 1, XLII.

53. As quoted in May, *Social Engineering*, 10.

54. As quoted in Gates, *Schoolbooks and Krags*, 114.

55. As quoted in Kramer, *Blood of Government*, 111. (Original is "Future Work in the Philippines," *New York Times*, February 7, 1899, 6.)

56. Kramer, *Blood of Government*, 112.

57. Miller, *Benevolent Assimilation*, 195.

58. Ibid., 188–189, 196–218.

59. As quoted in Gates, *Schoolbooks and Krags*, 114.

60. F. F Hilder, "Education in the Philippines," Annual Reports of the Department of the Interior. Report of the Commissioner of Education, 1898–1899, LX.

61. 1898–99 Report, vol. 2, 1621–1622; Kramer, *Blood of Government*, 42–101.

62. 1898–99 Report; Alexander A. Calata, "The Role of Education in Americanizing Filipinos," in *Mixed Blessing: The Impact of the American Colonial Experience on Politics and Society in the Philippines*, ed. Hazel M. McFerson (Westport, CT: Greenwood Press, 2002), 89–97; May, *Social Engineering*, 78.

63. Reports of the Taft Philippine Commission, 1900, 106–110.

64. May, *Social Engineering*, 79.

65. 1898–99 Report, 1631–1633.

66. Calata, "Role of Education," 91; May, *Social Engineering*, 85–89.

67. May, *Social Engineering*, chapter 5.

68. Kramer, *Blood of Government*, 116–120.

69. For an overview, see Go, "Introduction: Global Perspectives on the U.S. Colonial State in the Philippines." in *The American Colonial State in the Philippines*, ed. Julian Go and Anne L. Foster (Durham, NC: Duke University Press, 2003): 1–42.

70. Paul Kramer, "Empires, Exceptions, and Anglo-Saxons: Race and Rule between the British and U.S. Empires, 1880–1910," in *The American Colonial State in the Philippines*, ed. Julian Go and Anne L. Foster (Durham, NC: Duke University Press, 2003), 43–91.
71. Donna J. Amoroso, "Inheriting the 'Moro Problem' Muslim Authority and Colonial Rule in British Malaya and the Philippines," in *The American Colonial State in the Philippines*, ed. Julian Go and Anne L. Foster (Durham, NC: Duke University Press, 2003), 118–147.
72. Vince Boudreau, "Methods of Domination and Modes of Resistance: The U.S. Colonial State and Philippine Mobilization in Comparative Perspective," in *The American Colonial State in the Philippines*, ed. Julian Go and Anne L. Foster (Durham, NC: Duke University Press, 2003), 256–290.
73. May, *Social Engineering*, 123.
74. Calata, "Role of Education," 91.
75. May, *Social Engineering*, 124–125.
76. Kramer, *Blood of Government*, 347–431.
77. Boudreau, "Methods of Domination," 262; Miller, *Benevolent Assimilation*, 219–252.

Bibliography

A. B. A. "Education and Crime." *The School Review* 8, no. 1 (1900): 42–45.

Amoroso, Donna J. "Inheriting the 'Moro Problem' Muslim Authority and Colonial Rule in British Malaya and the Philippines," in *The American Colonial State in the Philippines*, ed. Julian Go and Anne L. Foster, 118–147. Durham, NC: Duke University Press, 2003.

Anderson, James D. "Education for Servitude: The Social Purposes of Schooling in the Black South, 1870–1930." PhD. diss., University of Illinois at Urbana Champaign, 1973.

Bacevich, Andrew. *American Empire: The Realities and Consequences of U.S. Diplomacy*. Cambridge, MA: Harvard University Press, 2004.

Bender, Thomas. *A Nation Among Nations: America's Place in World History*. New York: Hill and Wang, 2006.

Berthoff, Rowland. "A Country Open for Neighborhood." *Indiana Magazine of History* 84, no. 1 (1988): 109–126.

Bond, Jr., Beverly W. "American Civilization Comes to the Old Northwest." *The Mississippi Valley Historical Review* 19, no. 1 (1932): 3–29.

Boudreau, Vince. "Methods of Domination and Modes of Resistance: The U.S. Colonial State and Philippine Mobilization in Comparative Perspective," in *The American Colonial State in the Philippines*, ed. Julian Go and Anne L. Foster, 256–290. Durham, NC: Duke University Press, 2003.

Bloch, Ruth H. "Battling Infidelity, Heathenism and Licentiousness: New England Missions on the Post-Revolutionary Frontier, 1792–1805," in *The Northwest Ordinance: Essays on its Formulation, Provisions, and Legacy*, ed. Frederick D. Williams, 39–60. East Lansing, MI: Michigan State University Press, 1989.

Butchart, Ronald. *Northern Schools, Southern Blacks, and Reconstruction: Freemen's Education, 1862–1875*. Westport, CT: Greenwood Press, 1980.

Calloway, Colin. *The Scratch of a Pen: 1763 and the Transformation of North America*. New York: Oxford University Press, 2006.

Canny, Nicholas P. "The Ideology of English Colonization: From Ireland to America," in *Theories of Empire 1450–1800*, ed. David Armitage, 179–202. Brookfield, VT: Ashgate Publishing, 1998.

Carnoy, Martin. *Education as Cultural Imperialism*. New York: D. McKay Co., 1974.

Calata, Alexander A. "The Role of Education in Americanizing Filipinos," in *Mixed Blessing: The Impact of the American Colonial Experience on Politics and Society in the Philippines*, ed. Hazel M. McFerson, 89–97. Westport, CT: Greenwood Press, 2002.

Cogley, Richard W. *John Elliot's Mission to the Indians before King Philips War*. Cambridge, MA: Harvard University Press, 1999.

Cremin, Lawrence. *American Education Series*. New York: Harper and Row, 1970, 1980, 1988.

———. *American Education: The National Experience 1783–1876*. New York: Harper and Row, 1980.

Dolan, Jay. *American Catholic Parish: A History from 1850 to the Present*. New York: Paulist Press, 1987.

Elliott, John H. "The Seizure of Overseas Territories by the European Powers," in *Theories of Empire 1450–1800*, ed. David Armitage, 139–158. Brookfield, VT: Ashgate Publishing, 1998.

Elson, Ruth Miller. *Guardians of Tradition: American Schoolbooks in the Nineteenth Century*. Lincoln, NE: University of Nebraska Press, 1964.

Gates, John Morgan. *Schoolbooks and Krags: The United States Army in the Philippines, 1898–1902*. Westport, CT: Greenwood Press, 1973.

Go, Julian. "Introduction: Global Perspectives on the U.S. Colonial State in the Philippines," in *The American Colonial State in the Philippines*, ed. Julian Go and Anne L. Foster, 1–42. Durham, NC: Duke University Press, 2003.

Justice, Benjamin. "Beyond Nationalism: The Founding Fathers and Educational Universalism in the Early Republic," in *Advancing Democracy Through Education? US Influence Abroad and Domestic Practices*, ed. Bradley Levinson and Doyle Stevick. Information Age Publishing: Charlotte, 2008.

Kaestle, Carl F. "Public Education in the Old Northwest: Necessary to Good Government and the Happiness of Mankind." *Indiana Magazine of History* 84, no.1 (1988): 60–74.

Karnow, Stanley. *In Our Image: America's Empire in the Philippines*. Random House: New York, 1989.

Katz, Michael. *The Irony of Early School Reform*. Cambridge, MA: Harvard University Press, 1968.

Kellaway, William. *The New England Company 1649–1776: Missionary Society to the Indians*. New York: Longmans, 1961.

Kilian, Norbert. "New Wine in Old Skins? American Definitions of Empire and the Emergence of a New Concept," in *Theories of Empire 1450–1800*, ed. David Armitage, 307–324. Brookfield, VT: Ashgate Publishing, 1998.

Kramer, Paul. *The Blood of Government: Race, Empire, the United States, and the Philippines*. Chapel Hill, NC: University of North Carolina Press, 2006.

———. "Empires, Exceptions, and Anglo-Saxons: Race and Rule between the British and U.S. Empires, 1880–1910," in *The American Colonial State in the Philippines*, ed. Julian Go and Anne L. Foster, 43–91. Durham, NC: Duke University Press, 2003.

Love, Eric. *Race Over Empire*. Chapel Hill, NC: University of North Carolina Press, 2004.

May, Glenn Anthony. *Social Engineering in the Philippines: The Aims, Execution, and Impact of American Colonial Policy, 1900–1913*. Westport, CT: Greenwood Press, 1980.

McGreevy, John T. *Catholicism and American Freedom: A History*. New York: Norton, 2003.

Merrell, James H. "The Customs of Our Countrey: Indians and Colonists in Early America," in *Strangers Within the Realm: Cultural Margins of the First British Empire*, ed. Bernard Bailyn and Philip D. Morgan, 117–156. Chapel Hill, NC: University of North Carolina Press, 1991.

Miller, Stuart Creighton. *Benevolent Assimilation: The American Conquest of the Philippines, 1899–1903*. New Haven, CT: Yale University Press, 1982.

Monaghan, Jennifer. *Learning to Read and Write in Colonial America*. Amherst, MA: University of Massachusetts Press; Worcester: American Antiquarian Society, 2005.

Morris, Robert. *Reading, 'Riting, and Reconstruction: The Education of Freedmen in the South, 1861–1870*. Chicago, IL: University of Chicago Press, 1981.

Morrison, Dane. *A Praying People: Massachusett Acculturation and the Failure of the Puritan Mission, 1600–1690*. New York: Peter Lang, 1995.

Motyl, Alexander J. *Imperial Ends: The Decay, Collapse, and Revival of Empires*. New York: Columbia University Press, 2001.

Negri, Antonio and Michael Hardt. *Empire*. Cambridge, MA: Harvard University Press, 2000.

Onuf, Peter S. "Liberty, Development, and Union: Visions of the West in the 1780s." *The William and Mary Quarterly Third Series* 43, no. 2 (1986): 179–213.

———. *Statehood and Union: A History of the Northwest Ordinance*. Bloomington, IN: Indiana University Press, 1987.

Pomeroy, William J. *American Neocolonialism: Its Emergence in the Philippines and Asia*. New York: International Publishers, 1970.

Salisbury, Neal. "Red Puritans: The 'Praying Indians' of Massachusetts Bay and John Eliot." *The William and Mary Quarterly Third Series* 31, no. 1 (1974): 27–54.

Schirmer, Daniel B. and Stephen Rosskamm Shalom, eds. *The Philippines Reader: A History of Colonialism, Neocolonialism, Dictatorship and Resistance*. Boston: South End Press, 1987.

Sinopoli, Carla M. and Lars Fogelman, eds. *Imperial Imaginings: The Dean C. Worcester photographic collection of the Philippines, 1890–1913*. Ann Arbor, MI: University of Michigan Press, 1998.

Tanie, Norman Earl. "Education in John Elliot's Indian Utopias, 1646–1675." *History of Education Quarterly* 10 (1970): 308–323.

Tully, James. "Aboriginal Property and Western Theory: Recovering a Middle Ground," in *Theories of Empire 1450–1800*, ed. David Armitage, 345–372. Brookfield, VT: Ashgate Publishing, 1998.

Tyack, David. "Monuments Between Covers: The Politics of Textbooks." *American Behavioral Scientist* 42, no. 6 (1999): 922–932.

———. *Seeking Common Ground: Public Schools in a Diverse Society*. Cambridge, MA: Harvard University Press: 2004.

Tyack, David, Thomas James, and Aaron Benevot. *Law and the Shaping of Public Education, 1785–1954*. Madison, WI: University of Wisconsin Press, 1987.

Williams, Frederick D., ed. *The Northwest Ordinance: Essays on its Formulation, Provisions, and Legacy*. East Lansing, MI: Michigan State University Press, 1989.

Williams, Heather. *Self Taught: African American Education in Slavery and Freedom*. Chapel Hill, NC: University of North Carolina Press, 2005.

CHAPTER THREE

"The Path of Progress": Protestant Missions, Education, and U.S. Hegemony in the "New Cuba," 1898–1940

JASON M. YAREMKO

In the wake of the Spanish-Cuban-American war in 1898, Protestant missions from the United States landed in Cuba to deliver the word of the Gospel, combat the old colonial Catholic Church, and help "civilize" Cubans. Postwar Cuba attracted various North American interests. However, eastern Cuba, known as the "cradle of independence," was seen by both religious and secular U.S. interests as a "virgin field": historically isolated and little-influenced by either religious or secular corporations. While Protestant missionaries organized and erected churches, they also established various types of philanthropic institutions such as hospitals, schools, and colleges. Spreading the word of the Gospel—traditionally the principal evangelical approach—was not the only means for combating blasphemy and reaching individuals. Protestant missions in early republican Cuba adopted several methods of evangelization.[1] Of all these, U.S. churches in the newly-opened mission frontier of eastern Cuba, put the greatest emphasis on education.

Sunday schools, day schools, and colleges, missionaries surmised, would draw in Cubans for whom traditional preaching had no appeal. "New ideas" would be "patiently implanted" by missionaries like American Baptist Robert Routledge, the Canadian-born director of Baptist education in eastern Cuba, who insisted on the importance of student enrollment as an "opportunity for evangelization."[2] Although

Protestant schooling was initially a secondary evangelical tool, mission education programs in the eastern provinces soon outgrew their limited roles as agencies of evangelization. The attention to education was a tacit acknowledgment of the Social Gospel movement's influence on missions in Cuba.[3]

A facilitative institution in this respect was North American public education. Protestant missions and the U.S. government shared a common concern to direct education in Cuba. One of the first acts of the U.S. military government (1898–1902) was a decree for the construction of a new school system over the war ruins of the old.[4] As early as October 1898, Gilbert Harroun, secretary of the Cuban Education Association (CEA) and a founder of public education in Cuba, recruited teachers as part of the larger effort to "stamp the American educational system upon Cuban ignorance and laxity."[5] The program took off under General Leonard Wood, who, as governor of Oriente, established 200 schools. With Wood's assumption of the position of governor-general in 1900, some 3,000 schools were established in the first year. By 1902, over 250,000 Cuban children were in public schools.[6]

Protestant missions had no qualm about the U.S. domination of education in Cuba. Most missionaries had hoped that the growth of U.S. political and economic influence in Cuba would facilitate the penetration of "American and Christian ideals."[7] U.S. officials believed that Protestant schools had an important role to play and moved to encourage them "by every legitimate means."[8] Subsequent state and Protestant mission efforts at education in Cuba proved complementary. In a joint program with the CEA, for example, hundreds of Cubans "of loyal purposes and a zeal for education" were sent for training to colleges across the United States; many of these institutions were, like Earlham College in Indiana, founded and run by Protestant church organizations.[9] In turn, Cuban teachers (predominantly white) trained in these colleges were recruited from the CEA joint program by Protestant missions in Cuba.[10]

The slow development and inadequacies of the public school system reinforced the need for mission schools. By 1902, the U.S.-installed education system was considered a qualified success: the ranks of Cuban teachers surpassed 3,000 men and women; virtually all had taken part in a U.S. educational program. The first republican generation of public school children, notes Louis Pérez, received instruction from teachers trained in the United States. Yet rural areas still lacked schools, while urban primary schools were inadequate, and secondary education "was unknown in all of Oriente Province at that time."[11]

Missionary sentiment concluded that the combination of rampant illiteracy, the legacy of Spanish colonial neglect and "Romish oppression," and the abortive nature of the U.S. public education program, reinforced the great need for mission schools.[12]

The Rise of Protestant Mission Education

Protestant missions wasted no time in establishing new schools for training future leaders in mission pulpits, schools, homes, and society in general. By 1910, the three principle eastern missions accounted for dozens of schools, educating thousands of students enrolled in mission day, boarding and Sunday schools and offering curriculum at the primary and secondary levels. Many mission schools like the Baptist International Colleges in El Cristo became some of the most prestigious education centers in Cuba and received support from the business and governing classes. Despite some resistance from local Catholic priests, mission schools attracted overwhelming numbers of Cubans from all classes and in various regions. According to Baptist Howard Grose in 1908, openly avowed mission intentions did not dampen Cuban receptivity: "It was frankly announced that this was to be an evangelical school, that Christian influences would prevail, and that the Bible would be taught. Far from this deterring students from entering, the applications were so many as to be embarrassing to the management."[13] Among the more prominent institutions of Protestant education in eastern Cuba (and Cuba generally), were the Southern Methodists' Pinson Institute in Bartle and the Colegio Ingles in Camagüey, the famous Baptist International Colleges, and the Friends schools in Holguín, Gibara, and Banes.

The Protestant schools' most dynamic growth came during the first two decades of mission penetration. By about 1910, mission schools had expanded and diversified. Protestant education in eastern Cuba began with the establishment of day and Sunday schools. The rapid growth of these institutions, despite the vicissitudes of economic depression and sociopolitical unrest, persisted well into the second decade. Numerous mission stations reported class enrollments "booming" while others lamented the need to turn away dozens of students for a lack of capacity to accommodate their overwhelming success. The schools offered programs with the traditional curriculum, while a few also ran programs for specializing vocational and commercial training in agriculture, industry, and trades.[14]

Prestigious schools like the International Colleges in El Cristo, Pinson in Camagüey, and Los Amigos in Holguín were also boarding schools. While mission schoolrooms were filled by Cuban students of all classes, boarders, those who could afford to pay tuition, were predominantly of the middle class, and other families among the political and economic elite. Protestant missions looked to this latter group of students as the future leadership not only of the church but of the country as well, though "Christian education" oriented toward conversion and salvation had long been the primary function.

While Protestant schools in eastern Cuba responded to the educational needs of an impoverished postwar republic and an inadequate and ill-equipped public school system, they also served the larger evangelical concerns of the new Protestantism in Cuba. Furthermore, many among the mission's leadership privately admitted to the limits of religious work as an instrument of evangelization. After ten years, Protestant church membership, though steady in growth, had slowed relative to Protestant school numbers, which had caught up to, surpassed, and later dwarfed congregation numbers.[15] By the beginning of the second decade, education became the principle instrument for reinforcing the popular foundation of mission work of both evangelization and the formation of a Cuban ministry. Education became an essential medium for equipping Protestant churches and schools with future indigenous leaders. This was a long-term process, the ultimate goal of which, wrote one Protestant teacher, was the establishment of "a self-supporting, self-directing, self-propagating native church."[16] For the inculcation of values, beliefs, and conviction considered necessary in any dedicated believer, convert, or, ultimately, church leader, Protestant missionaries placed enormous faith in the Sunday school as the vehicle for evangelization and in children as the promise of Protestantism' success in Cuba.

Seen as future leaders of both ecclesiastical and secular Cuban society, children assumed the role as vessels of mission teachings, becoming instruments of evangelization at two significant levels. The first concerned the intrinsic advantage of teaching children. Cuban children, a Baptist missionary typically reported, "are easily molded and make rapid progress under wise and efficient teaching and discipline."[17] "The hope of the pioneer Church," another missionary concurred, "is in the children whose habits of life are not yet formed."[18] Children thus assumed an importance not only as malleable receptacles of North American Protestant beliefs and values, but also as leaders of the future Cuban Church.[19]

As mission students, children also served as educators, conduits through which Protestant mission influence reached Cuban adults, in many cases, successfully. American Baptist Gilbert N. Brink asserted that the "most hopeful point of contact is through the mission schools; by reaching the children the parent is reached also."[20] Mission schools therefore enabled missionaries to evangelize "to reach the children and through them the parents and thus do a much more effective work."[21] Religious study comprised a part of the curriculum at all levels of Protestant education. But the Sunday schools, seconded by the day schools, formed the foundation for evangelization and for mission schooling in general.

The Importance of Sunday Schools

The Sunday school was "the most efficient handmaiden of the church," according to at least one missionary.[22] By the 1930s, the Sunday schools of the Baptist, Methodist, and Friends missions in eastern Cuba recorded student enrollments approaching 10,000.[23] Sunday schools represented such an important institution for the inculcation of North American evangelical Protestant beliefs and values that they became an integral element in the mission consensus on education. The manifestation of such multi-denominational unity was the National Sunday School Association, which was administered by the North American members of the various missions, who looked upon the Sunday school as one important example of institutionalized benevolence.

Yet, as historians Thomas W. Laqueur and E.P. Thompson have demonstrated, if Sunday schools were a key component of "beneficent education," they also served as instruments of social control. Consciously or not, Sunday schools fostered conformity at the same time that they provided a form of moral and spiritual uplift, especially to those children for whom this mode of education was the only schooling they received. U.S. missionaries in eastern Cuba, just as their European counterparts, conceptualized the Sunday school as "an instrument for the moral rescue of poor children from their corrupt parents...and the regeneration of society."[24] That children were seen as "the advance troops, leading an invasion of godliness into their parents' houses" is as applicable to Cuba as to Europe.[25] The evangelical belief in the primacy of scripture and salvation was the principal motivation for the operation of Sunday schools, but not the only one.

Protestant Sunday schools also possessed a "civilizing" function not at variance with the needs of Cuba's foreign-dominated economy. Laqueur described one Sunday schoolteacher in Eighteenth-century England who suggested to potential subscribers that the "immoralities of the poor which keep their employers in a constant state of suspicion and uneasiness would be a thing of the past now that, because of her school, the education of poor children is no longer left entirely to their ignorant and corrupt parents."[26] Such sentiment endured several generations of Protestant Church development and had near-universal application, including mission fields in countries like Cuba. The civilizing component was typically expressed by a Baptist director of education in eastern Cuba who, in addition to noting the spiritual imperative of mission schooling, concluded that "Every converted man also at once becomes a reliable laborer and his services are preferred by the neighboring planters."[27]

Mission schools played a significant role in the inculcation of belief systems deemed by missionaries to be consistent with correct personal behavior. Yet as Laqueur correctly concluded, the political role of mission education varied over time and between cultures. As the cradle of independence, eastern Cuba was to prove a mixed success as a mission field. Conversely, however, while the eastern provinces' revolutionary origins might qualify the impact of Protestant mission education, they were at the same time the function of contradictions generated by U.S. hegemony and the missions' practical association therein.

If the ideological influence of the Sunday school was incomplete, its impact was significant. Protestant Sunday schools exceeded their churches in membership and development: compared to church growth, mission Sunday schools reportedly grew by "leaps and bounds," becoming one of the strongest Protestant institutions in organization and growth over and above that of the churches.[28] Sunday school growth was partly manifest in the increasing pleas made by missionaries for more school facilities, requests often generously met by private donors.[29]

As an educational institution, Protestant Sunday schools increasingly became the single greatest source for the evangelization of Cuban children. The Southern Methodist mission reported in 1925 that their Sunday schools accounted for at least 35 percent of Cubans converted to Methodist faith.[30] American Baptist Sunday school enrollment was more than double the Methodists, their proportion of converts perhaps significantly higher still. Logically, the overall impact of this form of religious education was limited because of the brevity (once a week) of exposure. However, this was only partly true as a generalization and even less applicable in the case of Cuba. In the context of mission education in eastern Cuba, it was not uncommon for day school teachers also to teach Sunday school; nor was it unusual for many day school students to attend Sunday school. As Baptist Charles White observed: "The general indifference on the part of the parents to Christianity...does not interfere with their desire to have their children well-educated, or cause them to hesitate to allow these same children to attend Sunday school and to be under the influence of the same teachers who give them instruction during the weekdays."[31] In some Southern Methodist day schools, the majority of students also attended Sunday school. Notably, many Cubans welcomed Protestant educational institutions at the same time that they resisted formal membership in the faith. As the Baptist mission's experience typically demonstrated, however, education was not neutral. "Some of the young people thus prevented in other years," one report concluded, "have later made a good profession of the faith gained in their childhood."[32] It was in this mutually-reinforcing sense that mission Sunday schools played a more significant role than they might have otherwise.

The Alignment of Mission Education with Civic and Economic Interests

Another part of the basis for the zeal with which mission schools were founded in Cuba was the perceived need for nurturing and molding a people consistent with the needs of a growing republic. As Baptist Herbert Grose asserted, Protestant education was crucial "also for those who are going to lead in public affairs and in business."[33] Secular interests in eastern Cuba were quite aware of the role played by mission schools. To missions and secular interests alike, each political protest, every subsequent U.S. intervention, reaffirmed the need for US tutelage in establishing stability and order in the Cuban republic. In turn, both parties believed that stability was necessary for good government and commerce. Popular rebellions in 1906 and 1912 renewed interest in the reinforcement of education generally, and Protestant education particularly, as an instrument for the dissemination of values, beliefs and practices consistent with a more stable republic.

After the second U.S. intervention in 1906, former U.S. Superintendent of Public Education, Alexis Frye, had advocated an adult education program expressly designed to accommodate both ecclesiastical and secular interests' concerns for stability: adult education was introduced to combat both Cuban illiteracy and, Frye argued, the apparent Cuban propensity for insurrection. Frye asserted that nearly all Cubans, "including the negroes who do most of the fighting," could be dissuaded from the insurrectionary habit and taught the histories of war's devastation in Europe, for example.[34] Deemphasizing socioeconomic conditions of underdevelopment as reasons for rebellions and subsequent U.S. interventions, Frye focused on the purported political immaturity of Cubans and proposed the establishment of adult programs and classes in towns and cities, especially in plantations, vegas, and villages. Adult education's purpose was in "making insurrection impossible among the present generation of laborers; for the future he would rely upon the Cuban children who had been instructed in the public and private schools of the island."[35]

Mission education policy identified very closely with the business and governing classes' needs concerning political and economic stability. Protestant schools were characteristically located in close proximity to urban centers, on or near plantations, vegas, mills, and company towns, as prescribed by Frye. Friends schools were typically located on the properties of North American companies like United Fruit and Chaparra Sugar Company in towns, such as, Banes, Chaparra, and Santa Lucia. Baptist and Methodist schools were also similarly centered in major areas of North American capital penetration like the Nipe Bay and Santiago regions where mill towns predominated.[36]

North American capital contributed substantially to the financing, construction, equipping, and, in numerous cases, staffing of Protestant schools in the early period of mission activity. The evolving relationship with mission schools assumed new forms and took on new supporters by the republic's second decade and beyond. The schools continued to expand as did the need for new facilities to meet the demand of company towns.[37]

Protestant schools were in fact popular for several reasons: They were "U.S. schools," which implied quality instruction; U.S. curriculum aided entrance into schools in the United States; and many courses were in English, widely thought to guarantee employment in a country increasingly geared toward U.S. corporations and tourists.[38] Finally, Protestant schools proved more stable than public schools, and more frequently demonstrated a vitality and resilience during politico-economic crises like the collapse of the Cuban "Dance of the Millions."[39] Throughout the first three decades of mission activity in eastern Cuba, Protestant schools were well-attended and more financially self-supporting than Protestant churches ever became. They trained thousands of Cuban students, the bulk of whom were not Protestants, but many of whom were or later became members of the economic and political elite.

Missionary educators persevered in their emphasis on the need to guide Cuba's rapidly developing public, social, economic and religious life within the framework of Protestant education. The teaching of arts, letters, and sciences, missionaries generally agreed, was aided by imposing a Protestant stamp on Cubans' moral and religious character. Sunday school and other educational literature in the form of text-books and teachers' manuals, whether published in the United States, or, later, in Cuba, remained predominantly North American in conception, even when some allowance was made for addressing local Cuban culture. The North American value system, often alluded to as the "universal experience of the race," was more often part of the "Anglo Saxon lesson of labor and thrift" oriented also toward "a desire for better things."[40]

In an educational environment of conservative, anti-Romanist, North American conceptions of Cubans' needs, mission educators taught their Cuban students how to be "good Christians," and "useful citizens."[41] Missionaries believed that all classes of Cubans benefited from exposure to Protestant conceptions of spiritual and material bet-terment.[42] Still, "universal morality" was not infrequently associated with North American, middle class values, or "Boston manners," as one Baptist missionary summarized it.[43] Missionaries, furthermore, did not view all Cubans as equally capable of leadership in religious or secular society. It is true that all classes of Cubans attended Protestant schools in eastern Cuba, but not all Cubans attended the same kinds of schools or classes. And missions favored some classes of Cubans over others.

Protestant mission schools and programs possessed a cultural and class bias—a function of North American Protestant education phi-losophy and policy generally, and also of the needs of Cuba's business and governing classes. While mission reports emphasized the need to establish schools for Cuban children from "the best families" as well as for those from "very poor homes," emphasis was placed on the need to educate and influence the youths of political and economic elites.[44] As one Southern Methodist report concluded, "the need for proper training for our ministry and for Christian teachers has been constantly before us; but we have also the responsibility for Christian professional and businessmen."[45] In the missions' perennial struggle to finance the expansion of Protestant schools, the rejection of growing numbers of students due to insufficient facilities was always problematic. Turning away children "from the influential homes of Cuba" was particularly agonizing: "Their fathers are the leaders of today and tomorrow. Turning these students away from their one opportunity to secure a Christian education means incalculable loss to the Kingdom of God."[46] Another was more to the point: "These are not charity pupils, but children whose parents are able and willing to pay.... The most urgent need of the Cuban mission is to provide the future leadership of our churches and of the island."[47]

Mission schools thus had several roles: The preparation of a Cuban ministry for the national Protestant Church of the future, the evangelization of the Cuban masses, and the education of Cuba's future business and governing classes—which included some members of Cuba's small middle class. In this latter role, the missions were perhaps the most successful. Missionary educators envisioned Protestant students from prominent families who "will one day be among the doctors, lawyers, planters and businessmen of the country, leading citizens in thought and action: They cannot help but take the influence of the school with them."[48] The rising interest of Cuba's affluent classes in mission schools was reflected in an equally growing attendance and increasingly active support for Protestant education. By 1920, the business–mission relationship had evolved beyond the beginnings of North American capital's financial and material aid for mission structures. National business and political leaders exercised an increasing influence over missionaries and mission policy, especially reflected in Protestant education. The fundamental values and interests of Protestant missions, and of the dominant political and economic institutions of Cuba's dependent economy and society, became mutually reinforcing.

The appeal of Protestant education drew many among the upper echelons of Cuban society, from national political and economic elites to local prestige groups at the municipal level. In addition to the patronage of North American capital, many Cuban political elites like first President Tomás Estrada Palma had long supported mission education. With the passing of the decades and of each political and fiscal crisis, support from these quarters increased and assumed various forms: official recognition, property concessions, military protection during uprisings, and perhaps most importantly, in their children's attendance at Protestant schools.

Protestant missions were quite conscious of attracting students from elite families and others among the middle class. These were, after all, groups which financially undergirded Protestant institutions and were to become Cuba's future church and national leaders. As mission reports put it, these were students "who are able to pay a good price for tuition and board and who have the promise of becoming leading citizens of the island."[49] By 1920, schools like the Baptist International Colleges had gathered prestige and won the support of governors, diplomats, presidents, vice presidents, and even public school superintendents, among others considered the "very best" people. Baptist mission reports boasted of the shining endorsements given their schools by a range of government officials that included everyone from Washington diplomat Carlos M. de Cespedes to President Menocal and his family.[50] The governor of Camagüey province gave both his public endorsement and children to the Methodist Pinson school, while American Friends educators boasted a list of families of "above average" Holguín school students that included the mayor of Holguín, state representatives, and the Secretary of the Board of Education.[51]

Many missionaries actively sought out government support and endorsement partly, as in most other Latin American countries, for school construction and operation, but also to gain accreditation and prestige.[52] Baptist, Methodist, and Friends schools were given official recognition and recommended by Cuban presidents and vice presidents, several of whom had also been presidents and managers of U.S. subsidiary companies in eastern Cuba.[53] Tomás Estrada Palma, Mario G. Menocal, and Alfredo Zayas were among the more powerful patrons.

However, not all mission reports of relations with elites were positive. Some among the Cuban political and economic elite—especially what remained of the Spanish large landholding class—continued to send their children to Catholic schools. Others employed tutors. Still other elites supported neither Roman Catholic nor Protestant institutions, but demanded the services of missionary educators as a virtual right.[54] And while Protestant mission schools also received public endorsement from the lower levels of government, this was by no means always so, as a few town councils stubbornly resisted both Protestant mission establishment and national government sanction.[55] Most mission stations, however, appeared to have been, if not always enthusiastically supported by local townships, at least accommodated. Government support, at all levels, was most consistent in the towns where Protestant prestige schools were established.

Conversely, while missions solicited and received support from foreign and indigenous elites, they encountered another form of resistance from within their own ranks. Cuban pastors and mission teachers had long protested against a generally Americocentric mission policy that appeared reluctant to turn the rhetoric of Cubanization into implementation. Cuban mission schools, like the churches, remained subordinated to U.S. curriculum, U.S. conferences, and North American administrators. From the early 1900s through the 1930s, Cuban mission school teachers registered their grievances through petitions, leaving the church, and/or starting their own, independent schools, often taking students with them.[56] Amid missionary lamentations, and criticism, of such dissension, U.S. missionaries and churches generally retained an essentially ethnocentric and paternalistic position that, as late as the 1940s, viewed most Cubans as incapable of self-government.[57]

Not all Cubans supported Protestant mission education policy per se, but many of the most important and influential among national and local elite interests, from Coca Cola to United Fruit, from Estrada Palma to Fulgencio Batista, did much in their power to facilitate Protestant mission education. By the early 1920s, mission schools in eastern Cuba had established themselves as the choice for many Cubans among the middle classes and of the political and economic elite. Support from these groups evolved along a continuum, beginning with financial and material aid for school construction, to participation at several levels of Protestant education development, including active participation in the formation of mission education policy.

The wish expressed by some missionaries for an increase in the active support of North American capital was eventually realized in the realm of mission policy formation, as companies like United Fruit and Chaparra Sugar assumed roles in influential bodies such as committees for Protestant school construction and curricular development.[58] In May 1919, Southern Methodist missionaries in Camagüey conveyed their gratitude to the company managers, bankers, and other prominent businessmen who had headed one such school committee.[59]

As Protestant missions continued to depend on the substantial donations and support of North American capital and Cuban political elites for their schools' success, Protestant education programs increasingly came under the direct influence of these secular interests. By the 1920s, the demand for business schools and commercial programs was being thoroughly met by schools like the International Colleges, the Pinson school, and Los Amigos—the cream of the principal Protestant schools in eastern Cuba whose education programs were often subsidized by and conceived in conjunction with political and economic elites. Since at least 1910, when the Methodist Candler College in Havana acquiesced to the calls of prominent bankers and businessmen for a business school, numerous Protestant schools in eastern Cuba developed their own programs for business and commercial training. Candler had its eastern equivalents in the Protestant schools in El Cristo, Camagüey, and Holguín. All played substantive roles in preparing functionaries for North American business interests and the Cuban government. The network of graduates that emerged from these schools "not infrequently smoothed the way for companies having difficulties with the government."[60] All of these schools earned national reputations for excellence in education generally and for vocational training in particular.

Unequal Access to Educational Opportunities

Not all Cubans had access to the same Protestant schools and programs: "vocational training" had different meanings for different classes of Cubans. While Protestant missions declared in favor of the development of fraternity, equality, and of the general spiritual and material betterment of all Cubans, mission reports alluded to the structural divisions, which Protestant education programs increasingly reinforced. A 1914 report on American Baptist school work in Camagüey noted the rising attendance of students from the "best families" as well some students from impoverished households; the bulk of the poor students attended the Baptist Industrial School in Camagüey.[61]

Industrial or vocational education became a significant part of Protestant education programs in eastern Cuba. Scholars have argued that these forms of schooling were founded largely in order to train African Americans and lower class whites as laborers and domestic servants in order to meet the growing needs of an industrialized United States.[62] As North American business interests and Cuba's governing classes increasingly lent support to Protestant missions, the missions reciprocated by developing curricula, along with the technical and vocational courses and programs, desired by U.S. business interests: Cubans received training for employment with North American companies at all levels. The Preston school, for example, provided programs consistent with the needs of United Fruit, and mission teachers were paid by the company.[63] The levels of training provided by mission schools corresponded to the kinds of skills needed by North American enterprises, and, subsequently, with the social and racial divisions of Cuban society. In this manner, U.S. Protestant missions reinforced North American hegemony in eastern Cuba specifically, in a manner consistent with what occurred across Cuba generally.

While initially few, the number of Protestant schools geared solely toward industrial or technical training in eastern Cuba expanded by the 1930s and 1940s. As early as 1920, the Friends mission school in Holguín had already opened a commercial department with courses that included English, typewriting, and accounting. Missionaries asserted that the department was created because "there was a great demand for it, because it offered a means of replenishing the straightened school treasury, and because it would bring the missionaries in touch with a class of young people that would probably not otherwise be reached."[64] Friends' other principle schools in Gibara, Banes, and Puerto Padre were modeled after the Holguín school. Southern Methodist and American Baptist schools also developed and expanded the commercial content of their educational programs; the former running vocational programs in the Pinson school in Camagüey and Preston in Oriente, the latter operating an industrial school in Camagüey since at least 1910, while also pursuing further development of industrial and commercial education programs at the International Colleges in El Cristo.[65]

Missions' drive to meet the demand for industrial and mechanical training intensified with the boom in Cuba's prosperity during and after World War I.[66] These were years of "dazzling prosperity" in which sugar prices soared and production expanded, and climaxed in 1920, the year of the "dance of the millions." As the Cuban economy expanded from 1915 to 1920, so did the operations of North American banks and mills.[67] The expansion of North American capital in eastern Cuba intensified companies' demands on Protestant schools for more capitalist-oriented education programs, which the missions strived to meet.

Yet Protestant schools' increasingly business-oriented curricula were framed by contradictory goals. The promotion of a spiritual life and the free development of the potential of the individual conflicted with the other aim of providing practical skills and an ideology of corporate efficiency; short-term training programs clashed with missionary ideals of equality of opportunity.[68] Furthermore, despite missionaries claims that their schools' curricula addressed the children of the working and middle classes equally, mission schools and programs responded more "to the backgrounds and possibilities of the children of the white collar employees and businessmen rather than to those of the children of the workers."[69] The Protestant prestige schools of Los Amigos, Pinson, and International Colleges increasingly accommodated those who became the doctors, lawyers, and judges of eastern Cuba. The social aspirations of the Cuban working and lower classes were fostered by Protestant education programs less according to individual potential than to those classes' presumed station in life.

As Baptist Robert Routledge observed: "Each year that passes sees a larger class of graduates going out—some to prepare for professional life...some to teach in our schools, and others to enter the business life of the community."[70] Most missionaries regarded these kinds of graduates as the "best advertisement" for mission schools in attracting more of the kind of student who could pay tuition and board.[71] As the needs and expectations of educators and business interests continued to converge, contradictions in mission principles and practice became increasingly apparent.

Not unlike the Spanish colonial church whose educational institutions also distinguished between the elites and the *gente baja*, or lower classes, Protestant schools helped reinforce class divisions in Cuban society by means of discriminatory education programs.[72] Characteristically, though the classes may at times have mingled, children of the elite were more often paying boarders while the children of the working classes were invariably day students. Furthermore, the "charity pupils" stations in life seemed predetermined and thus reinforced by the limited opportunities afforded them by their education, especially evident when one notes the kinds of technical schools to which children of workers and the poor were restricted in comparison to their middle class counterparts. Protestant missions fostered industrial education as a means to combat the perceived indolence of the Cuban poor, and to enable them to learn the "dignity of manual labor," so as to become useful citizens' and good employees.[73] This included courses in subsistence gardening, wood and brick work, building repair, and other forms of manual labor. Friends day schools also held "handwork" classes; in the Gibara school, for example, the curriculum included paperwork, embroidery, and basket weaving.[74]

Protestant industrial education programs both fed on class divisions and were reflections of them. Missions' reinforcement of the social hierarchy in eastern Cuba was also an expression of one of the fundamental aspects of general Protestant mission practice, as conveyed in an exceptionally insightful observation by Baptist missionary George Rice Hovey, who revealed something of the universality of the mission experience in Cuba. In a 1921 mission education report, Hovey noted an essential point in common of the mission experience in Cuba, Puerto Rico, and among African Americans in the southern United States: the emergence of a two-tiered, class-based system in worship and education: "Where abject poverty prevails, the schools almost of necessity are giving courses in agricultural and industrial work."[75]

Protestant Missions and Gender, Race, and Nationalism in Cuba

In addition to reinforcing U.S. economic and political control in Cuba, Protestant mission schools also facilitated gender and race divisions. Women's education was oriented toward the improvement of women's social condition as well as their preparation for the labor market. Women mission groups like the Methodist Women's Missionary Society and the Women's American Baptist Home Mission Society ran numerous schools and courses in addition to the coeducational primary, secondary, and Sunday schools. These included girls' boarding schools and industrial schools managed in conjunction with other Protestant education programs in mission stations. American Baptist missions operated a Girls' Industrial School in towns throughout eastern Cuba, as did the Friends in the form of *casas hogares* or "domestic science" schools, which promoted the discipline and "dignity of the home."[76]

Consistent with George A. Coe's thesis on religious education, Protestant missions in Cuba perceived the family as the foundation of political democracy. Conceived in Protestant American bourgeois terms, this meant the nurturing of traditional ideals of women's roles. Mission education programs reinforced this direction in domestic training beyond Cuban women's school years and into adult life. This was done by forming various women's groups like the Baptist "What I Can" society and other philanthropic women's organizations. Sunday schools also included time to spend on embroidering or sewing. According to one woman missionary, "all Cuban women and girls love that kind of work."[77] Not all women's education focused on domestic training; missions also prepared Cuban women for the labor market. For the most part, this meant secretarial work for North American companies. It also included training as mission and public school teachers; few received training as missionaries. Women employed in schools and U.S. companies, however, were typically those who could afford to pay the tuition for that level of education. Protestant schools offered real advantages and opportunities for the middle class and others among the elite for whom public schooling was inadequate.

Protestant mission schools also offered solutions to the politics of race in Cuba. Many missionaries came from the southern United States—especially Methodists and Baptists—and arrived in eastern Cuba "with ways to address the concerns of white Cubans."[78] Missionaries recognized one of the chief concerns of white Cubans of race-mixing in public schools as a propitious opportunity for missions. Such conditions created "a demand for first-class private schools on the part of persons who are able to pay for the education of their children, thus being opened up a way of access for the missionary to a large and influential class of people."[79] Friends mission schools were not immune to the "necessity" of drawing the color line in order to attract the "better grade Cubans" who did not want their children "thrown in with colored children" in the public schools.[80] Subsequently, segregated education was no stranger to Protestant schools in early republican Cuba.

Yet Protestant education was more than the function of corporate interests and the Cuban economy. Demand for this type of education also came from the Cuban middle class, for whom the programs provided skills highly regarded as a way of maintaining middle class status.[81] This was also the group from whom missions expected much of Cuba's future leadership would derive. By the fourth decade of mission endeavor, some of the most prestigious business schools in Cuba were Protestant schools in the east.[82] Charles Detweiler aptly demonstrated one element of the relation of Protestant education to Cuban society in his 1923 description of a typical student's evolution:

A number of years ago a young man began to come to a mission Sunday School in Cuba.... Soon after entering the Sunday School he joined the Young People's Society and later took up special studies.... The Christian ideals of life gripped him with a compelling force. As his life flowed out to others there came to him a new appreciation of what is worthwhile in material things. The missionary found that he had all unconsciously created a new market for American goods.[83]

In spite of missions reports' emphasis on the imperative of conversion and Cuban ministry development, missionaries' other overriding concern lay with the nurturing of responsible Christian corporate citizens. National stability and prosperity remained an essential part of the evangelizing and "civilizing" or "moralizing" mission, which by the 1930s appeared to have confirmed its place as a means to the end of evangelization. This goes a long way to explain the increased emphasis on education among Protestant missions in eastern Cuba, as well as the enthusiasm for cooperation among missions in this area of endeavor. Finally, the U.S. churches' perceived need to direct and dominate education programs for the sake of church and country was also palpable in their negative reactions to indigenous demands for Cubanized administration as well as to the growing activism of nationalistic students. Cubans were more readily considered trainable as pastors and Protestant school teachers than as administrators; only rarely and much later did a select few eventually become superintendents. After four decades, Protestant missions boasted that most or all of their churches were under the charge of Cuban pastors. But as one missionary concluded earlier: "In our educational work, we have gone more slowly."[84] In civil society, Protestant churches reaffirmed their confidence in the structures and surrogates of U.S. hegemony, even as these were increasingly challenged by a growing number of Protestant school graduates like Frank País. Not unlike secular society in Cuba, Protestant education remained under the effective control of U.S. administrators, until Cuba's revolutionary nationalism changed everything.

Notes

Excerpts from Jason M. Yaremko, *U.S. Protestant Missions in Cuba: From Independence to Castro* (Gainesville, FL: University Press of Florida, 2000), reprinted with the permission of the University Press of Florida.

1. Kenton J. Clymer, *Protestant Missionaries in the Philippines, 1898–1916: An Inquiry into the American Colonial Mentality* (Urbana, IL: University of Illinois Press, 1986), 18.
2. Henry Morehouse, *Ten Years in Cuba* (New York: American Baptist Home Mission Society, 1910), 8–10, 12, 14; American Baptist Home Mission Society (ABHMS) *Annual Report of the Board*, 83rd Sess., 1915 (New York, 1915), 99–100; ABHMS *Annual Report*, 79th Sess., 1911 (New York, 1911), 39.
3. Robert Wauzzinski, *Between God and Gold: Protestant Evangelicalism and the Industrial Revolution, 1820–1914.* (Cranbury, NJ: Fairleigh Dickinson University Press, 1993).
4. See, Erwin H. Epstein, "The Peril of Paternalism: The Imposition of Education on Cuba by the United States," *American Journal of Education* 96, no. 1 (November 1987): 1–23.
5. Gilbert K. Harroun to General Joseph Wheeler, October 19, 1898, Cuban Education Association Papers, Manuscripts Division, Library of Congress, Washington, DC (hereafter MD/LOC).
6. Ada Ferrer, "Education and the Military Occupation of Cuba: American Hegemony and the Cuban Response, 1898–1909," (MA Thesis, University of Texas, Austin, 1988), 31–32, 44–68; Louis A. Pérez, Jr., "The Imperial Design: Politics and Pedagogy in Occupied Cuba, 1899–1902," *Cuban Studies* 12, no. 2 (1982): 7–10.
7. Charles Detweiler, *Twenty Years in Cuba* (New York: General Board of the Northern Baptists Convention, 1923), 3–4.
8. "Memo for Mr. Harroun Regarding Education System for Cuba," November 29, 1898, Cuban Education Association Papers, MD/LOC.
9. Gilbert Harroun to Manager of the Associated Press, November 11, 1898; J. Mills, President, Earlham College, to Gilbert Harroun, November 23, 1898, Cuban Education Association Papers, MD/LOC.
10. George Hodge, YMCA, to A.W. Raymond, February 8, 1899, Cuban Education Association Papers, MD/LOC.
11. Hiram Hilty, *Friends in Cuba* (Richmond, IN: Friends United Press, 1977), 48.
12. Howard E. Grose, *Advance in the Antilles* (New York: Eaton & Maine, 1910), 49–51.

13. Howard Grose, "The Cuba Trip," *Baptist Home Missions Monthly (BHMM)* 30, no. 4 (April 1908): 170.

14. For a fuller treatment, see Jason M. Yaremko, *U.S. Protestant Missions in Cuba: From Independence to Castro* (Gainesville, FL: University Press of Florida, 2000).

15. Charles L. White to Katherine J. Westfall, May 19, 1914, ABHMS Correspondence, Cuba, American Baptist Historical Society, Valley Forge, Pennsylvania (hereafter cited as ABHS); ABHMS *Annual Report*, 83rd Sess., 1915 (New York, 1915), 75–76, 94–100.

16. Clarence McClean, "Pioneer Work of the Friends in Cuba" (MA Thesis, University of Chicago, Chicago, 1918), 21, 23.

17. Grose, *Advance*, 111.

18. McClean, "Pioneer Work," 15–16.

19. Grose, *Advance*, 112–113.

20. ABHMS *Annual Report*, 83rd Sess., 1915 (New York, 1915), 75–76.

21. A. B. Howell to W. C. Treat, May 18, 1914, ABHMS Correspondence, Cuba, ABHS.

22. McClean, "Pioneer Work," 15.

23. MECS, *AC*, 1936 (Habana, 1936), 45–46; American Friends Board of Foreign Missions (AFBFM) *Annual Report*, 1936 (Richmond, 1936), 70–71; ABHMS *Annual Report*, 103rd Sess., 1936 (New York, 1936), 65.

24. Thomas H. Laqueur, *Religion and Respectability: Sunday Schools and Working Class Culture, 1780–1850* (London: Yale University Press, 1976), 6–8, 9–10.

25. Ibid.

26. Ibid.

27. Robert Routledge, "A Survey of the Eastern Cuban Mission," *Missions* (October 1930): 525.

28. AFBFM *Annual Report*, 1928 (Richmond, 1928), 56–57; ABHMS *Annual Report*, 89th Sess., 1921 (New York, 1921), 78–80.

29. ABHMS *Annual Report*, 89th Sess., 1921 (New York, 1921), 79–80; AFBFM *Annual Report*, 1930 (Richmond, 1930), 28–30.

30. MECS, *AC*, 1925 (Habana, 1925), 74.

31. Charles White to Westfall, May 19, 1914, ABHMS Correspondence, Cuba, ABHS; MECS, *AC*, 1921 (Habana, 1921), 39.

32. Charles White, "A Sunday Afternoon at Cristo," *Missions* (May 1915): 448–450.

33. Grose, *Advance*, 112.

34. Cited in Charles Elliot, President, Harvard University to William H. Taft, July 12, 1907, RG 199, NARA.

35. Ibid. Frye's plan was later rejected by Governor Charles Magoon and then Secretary of Public Education Lincoln de Zayas See Charles Magoon, Provisional Governor, to William H. Taft, September 4, 1907, RG 199, NARA.

36. "Memory Books of Sylvester and May Mather Jones, 1900–1960," 84–85, 91–92; Sylvester and May Mather Jones Papers, FHC; ABHMS *Annual Report*, 71st Sess., 1903 (New York, 1903), 111–114; MECS, *AC*, 1920 (Habana, 1920), 40, 74–75.

37. AFBFM *Annual Report*, 1919 (Richmond, 1919), 48–49; AFBFM *Annual Report*, 1930 (Richmond, 1930), 28–30; ABHMS *Annual Report*, 89th Sess., 1921 (New York, 1921), 78–80.

38. Crahan, "Religious Penetration and Nationalism in Cuba, 1898–1958," *Revista/Review Interamericana* 8, no. 2 (1978): 219–220.

39. AFBFM *Annual Report*, 1922 (Richmond, 1922), 17–24.

40. Clymer, *Protestant Missionaries in the Philippines, 1898–1916*, 86–87; Detweiler, *Twenty Years*, 27–28.

41. Sterling Neblett, "Religious Education," memorandum, January 1929, 1–11, Cuban Conference Correspondence, UMCA.

42. Detweiler, *Twenty Years*, 27–28.

72   *Jason M. Yaremko*

43. H.R. Moseley, "Young Cuba," *Missions* (April 1912): 364–367.
44. "School Work in Camagüey, Cuba," *Missions* (February 1914): 136–137.
45. MECS, *AC*, 1933 (Habana, 1933), 55.
46. "Shall We Turn Them Away?" *Missions* (February 1919): 109.
47. Charles S. Detweiler, "A Brief Survey of Our Latin American Missions," *Missions* (January 1921): 74.
48. James A. Stewart, "First Impressions at El Cristo," *Missions* (May 1930): 287; AFBFM, *Friends Missionary Advocate*, 1915 (Richmond, 1915), 19.
49. "Report on Colegio Internacionales (Cristo)" (1916), 1–3, ABHMS Correspondence, Cuba, ABHS.
50. Ibid. "Magnifico discurso" (Speech by Fernando García Grave de Peralta, Governor of Oriente province, at the Colegios Internacionales, El Cristo), El Mensajero (July 15, 1927), 8.
51. AFBFM *Annual Report*, 1923 (Richmond, 1923), 28–29.
52. Susan Martin to Eva and Samuel Haworth, August 29, 1916, Martin and Haworth Family Papers, FHC.
53. Zenas Martin to Susan Martin, April 22, 1902; Susan Martin to Eva and Samuel Haworth, February 28, 1917, Martin and Haworth Family Papers, FHC; Mario G. Menocal to Ezeqíel García, February 12, 1915, Fondo Donativos y Remisiones, Legajo 373, no. 20, ANC; President Alfredo Zayas to Bishop James Atkins, February 18, 1922; R. Martínez, Secretary of Gobernación to James Atkins, February 1922; A. Figueredo, General Treasurer, to James Atkins, February 1922; Ben O. Hill to Warren A. Candler, January 4, 1929, Warren Aiken Candler Papers, ESCD.
54. AFBFM *Annual Report*, 1921 (Richmond, 1921), 25.
55. See correspondence of V. Martínez and R. Herrera, Santiago de Cuba, October 13 to December 29, 1910, Fondo Gobierno Provincial, Iglesias, Legajo 770, no. 18, Archivo Historico Provincial, Santiago de Cuba (hereafter cited as AHPS).
56. Jason Yaremko, "Protestant Missions, Cuban Nationalism, and the Machadato," *The Americas* 56, no. 3 (January 2000): 64–66.
57. Ibid.
58. AFBFM *Annual Report*, 1918 (Richmond, 1918), 25–27.
59. Ben O. Hill to E.H. Rawlings, May 20, 1919, Cuban Conference Correspondence, UMCA.
60. Crahan, "Religious Penetration," 220.
61. "School Work in Camagüey, Cuba," *Missions* (February 1914): 136–137.
62. James Anderson, *Education of Blacks in the South, 1860–1935* (Chapel Hill, NC: University of North Carolina Press, 1988); Herb Kliebard, *Schooled to Work: Vocationalism and the American Curriculum 1876–1946* (New York: Teachers College Press, 1999); and, Donald Spivey, *Schooling for the New Slavery: Black Industrial Education, 1868–1915*. (Westport, CT: Greenwood Press, 1978).
63. Irene Wright, " The Nipe Bay Country—Cuba," *Bulletin of the Pan-American Union* (June 1911): 991.
64. AFBFM *Annual Report*, 1921 (Richmond, 1921), 25–26.
65. Actas de la Convención Bautista de Cuba Oriental (CBCO), *Estadisticas Anuales*, 1910, 70, 73–74.
66. ABHMS *Annual Report*, 77th Sess., 1909 (New York, 1909), 98; ABHMS *Annual Report*, 88th Sess., 1920 (New York, 1920), 45, 48–50.
67. Louis A. Pérez, Jr., *Cuba: Between Reform and Revolution* (New York: Oxford University Press, 1995), 224–226.
68. Rosa del Carmen Bruno-Jofré, *Methodist Education in Peru: Social Gospel, Politics, and American Ideological and Economic Penetration, 1888–1930* (Waterloo, ON: Wilfred Laurier University Press: 1988), 159.

69. Ibid., 168.
70. ABHMS *Annual Report*, 92nd Sess., 1925 (New York, 1925), 43.
71. ABHMS *Annual Report*, 90th Sess., 1922 (New York, 1922), 63.
72. Robert Ricard, *The Spiritual Conquest of Mexico: An Essay on the Evangelizing Methods of the Mendicant Orders in New Spain*, Trans. Lesley Byrd Simpson (London: University of California Press, 1966), 98–99, 207–209, 212–216.
73. Clymer, *Protestant Missionaries in the Philippines*, 86.
74. AFBFM *Annual Report*, 1917 (Richmond, 1917), 27–28; Memory Books of Sylvester and May Mather Jones, 1900–1960, 48–50, Sylvester and May Mather Jones Papers, FHC.
75. ABHMS *Annual Report*, 89th Sess., 1921 (New York, 1921), 61.
76. Elizabeth M. Allport; Kathleen Rounds, "To the Women's American Baptist Home Mission Society," Pamphlet, 1957, ABHS; Memory Books of Sylvester and May Mather Jones, 1900–1960, 264, Sylvester and May Mather Jones Papers, FHC.
77. Women's American Baptist Home Mission Society (WABHMS) *Annual Report*, 1915 (New York, 1915), 219.
78. Cited in Pérez, *Essays*, 67.
79. Ibid.
80. Cuban Field Committee to Zenas Martin, Holguin, April 1, 1909, Wider Ministry of Friends United Meeting Papers, Lilly Library, Earlham College, Richmond, Indiana (hereafter WMFUMP).
81. Bruno-Jofré, *Methodist Education in Peru*, 164.
82. Detweiler, *Twenty Years*, 27–28.
83. Ibid.
84. ABHMS *Annual Report*, 93rd Sess., 1926 (New York, 1926), 64.

Bibliography

Anderson, James. *Education of Blacks in the South, 1860–1935*. Chapel Hill, NC: University of North Carolina Press, 1988.

Bruno-Jofré, *Rosa del Carmen*. *Methodist Education in Peru: Social Gospel, Politics, and American Ideological and Economic Penetration, 1888–1930*. Waterloo, ON: Wilfred Laurier University Press: 1988.

Clymer, Kenton J. *Protestant Missionaries in the Philippines, 1898–1916: An Inquiry into the American Colonial Mentality*. Urbana, IL: University of Illinois Press, 1986.

Crahan, Margaret E. "Religious Penetration and Nationalism in Cuba, 1898–1958." *Revista/Review Interamericana* 8, no. 2 (1978): 204–224.

Epstein, Erwin H. "The Peril of Paternalism: The Imposition of Education on Cuba by the United States." *American Journal of Education* 96, no. 1 (1987): 1–23.

Ferrer, Ada. "Education and the Military Occupation of Cuba: American Hegemony and the Cuban Response, 1898–1909." MA Thesis, University of Texas, Austin, 1988.

Hilty, Hiram. *Friends in Cuba*. Richmond, IN: Friends United Press, 1977.

Kliebard, Herb. *Schooled to Work: Vocationalism and the American Curriculum 1876–1946*. New York: Teachers College Press, 1999.

Laqueur, Thomas H. *Religion and Respectability: Sunday Schools and Working Class Culture, 1780–1850*. London: Yale University Press, 1976.

McClean, Clarence. "Pioneer Work of the Friends in Cuba." MA Thesis, Chicago: University of Chicago, 1918.

Pérez, Jr., Louis A. "The Imperial Design: Politics and Pedagogy in Occupied Cuba, 1899–1902." *Cuban Studies* 12, no. 2 (1982): 1–19.

Ricard, Robert. *The Spiritual Conquest of Mexico: An Essay on the Evangelizing Methods of the Mendicant Orders in New Spain.* Trans. Lesley Byrd Simpson. London: University of California Press, 1966.

Spivey, Donald. *Schooling for the New Slavery: Black Industrial Education, 1868–1915.* Westport, CT: Greenwood Press, 1978.

Wauzzinski, Robert. *Between God and Gold: Protestant Evangelicalism and the Industrial Revolution, 1820–1914.* Cranbury, NJ: Fairleigh Dickinson University Press, 1993.

Yaremko, Jason M. *US Protestant Missions in Cuba: From Independence to Castro.* Gainesville, FL: University Press of Florida, 2000.

Yaremko, Jason. "Protestant Missions, Cuban Nationalism, and the Machadato." *The Americas* 56, no. 3 (2000): 53–75.

CHAPTER FOUR

American Philanthropy and Reconstruction in Europe after World War I: Bringing the West to Serbia

NOAH W. SOBE

Three American armies invaded Europe in the years of World War I and in its aftermath—at least such was the account proposed in 1924 by the Serbian Child Welfare Association of America. First came the American Expeditionary Force, which entered the war in 1917 after the European combatants had been fighting for three years. The second "American army" was the American Relief Force that arrived after the armistice of November 11, 1918; the third was the "Army of Reconstruction." And, according to the Serbian Child Welfare Association, "the first army helped to set Europe free; the second lifted her and set her on her feet; the third army started her on her way rejoicing toward a higher civilization."[1] As will become clear in this chapter, the activities of these three "armies" were not as clear-cut and distinct as portrayed here, nor were they necessarily separated and neatly sequenced, however it is not to be contested that during and after World War I a substantial number of Americans invaded Europe with notions of freedom, uplift, and civilization on their minds.

A striking feature of this U.S. reconstruction and relief work in Europe was its voluntary and philanthropic character. While the "doughboys" in General Pershing's Expeditionary Force were conscripted, they were supported by tens of thousands of individual American volunteers: ambulance drivers sent by dozens of American colleges and universities; over 20,000 "red triangle" YMCA workers who set up "Y huts" and commissaries to offer recreational and educational opportunities for American soldiers; as well as American Red Cross nurses and doctors who augmented the U.S. Medical Corps.[2] Alongside the charitable resources marshaled to support Americans fighting in Europe, the welfare of Europe's suffering civilian populations was a key American concern from the very outbreak of conflict. Notable initiatives include a 1915 medical mission to Serbia to combat typhus, Herbert Hoover's feeding programs in German-occupied Belgium, as well as a whole variety of projects to aid Europe's displaced and orphaned children—including, from 1917 on, an American Junior Red Cross that raised unexpectedly large sums from American school children.[3] Distress at the distant suffering of others was certainly central to these charitable initiatives. The purpose of this chapter is neither to valorize nor to fault Americans' humanitarian impulses, but rather to establish what else was carried to Europe as Americans came over to offer aid and reconstruct societies devastated by conflict. To the extent that these various projects defined the problems at hand, envisioned certain solutions (thus excluding others), and viewed some behaviors and subjectivities as "proper" and normal (again excluding others as "backward" or "uncivilized"), we can consider these voluntary charitable projects as projections of American power and influence.

Centuries-old projections of America as "new" world and Europe as "old" inevitably framed American philanthropic good works designed to alleviate suffering in Europe. American exceptionalism has been long been predicated on city-on-a-hill imagery suggesting that salvation and redemption is to be found in a world remade in the American image.[4] This logic is evident in many of the accounts of reconstruction activities discussed below. In the Serbian Child Welfare Association notion that American assistance could help Europeans rejoice "toward a higher civilization" there was a dramatic—if long anticipated— reversal of civilizing referents. In terms of discursive positioning, one might say that by the early twentieth century America had achieved enough "maturity" (or enough confidence in its parvenu status) to now civilize the old Europe that had traditionally claimed global preeminence as the privileged locus of civilization and modernity. Victoria de Grazia insightfully argues that America's global (hegemonic) prominence was built on European territory. It was the European validation of American authority for norms-making that enabled the representation of all that is advanced and modern to shift to America.[5] That there had been a transfer of certain forms of authority was remarked upon at the time and became fodder for "clash of civilizations" analyses (American *vs.* European) that has been a special preoccupation of European intellectuals up through the present day. Nonetheless, the 1914–18 war represented a moment of particular crisis for European self-definition. Peter Wagner argues that it is at this moment that the self-image of Europe as the seat of modernity was effectively replaced within Europe with a discourse of tradition.[6] In this chapter I propose that World War I and its aftermath provided one of the key moments for demonstrating America's modern superiority in the areas of child welfare, social welfare generally, and education.

The overseas philanthropic activity of American charitable foundations has been the subject of considerable academic attention, with scholarship tending to emphasize the period after 1945 when the institutions such as Ford, Carnegie, and Rockefeller supported scientific and social science research as well as policy reforms around the globe.[7] At least in the initial decades after World War II this work took place under the spell of modernization theory and was powerfully informed by Cold War politics.[8] However, while U.S. foundations certainly played a pivotal role on the international scene from the 1950s on, the international dimension of their work extends back into the interwar period. In fact, Rockefeller's international health division was principally active in the 1920s and 1930s, with the global management of health (both disease eradication and the development of local public health facilities) eventually coming under the auspices of UN agencies such as the World Health Organization from the late 1940s onward.[9] As Ellen Lagemann's work on the Carnegie Corporation has shown, one feature that endows foundation philanthropy with norm-setting influence is simultaneous engagement with both policy formation and the production of the scientific/scholarly knowledge that said policies are based on.[10] She argues that the split between sociology and social work can be seen in the work of American foundations, specifically in the early 1920s with Carnegie's emphasis on "scientific philanthropy" and support for expert decision-making and specialized, professionalized work in the social sciences—a trend that would take on considerable momentum after World War II. However, even if this represented a stark contrast to earlier models of social science research (as epitomized in the settlement house social survey),[11] it was very much the earlier, applied social work model that domi-nated American reconstruction projects in war-torn Europe. Thus, while post–WW I American philanthropic involvement overseas did not exactly anticipate the modernization problematic of later decades, it still operated as a norm-setting, knowledge-producing enterprise. Moreover, what we see at this moment in terms of the stance Americans took toward international work was quite con-sistent with approaches to international reconstruction work taken earlier and later. Salient here, as I will argue, is the ease with which Americans could see themselves as justly and unproblematically intervening in public spheres outside the United States.

Rockefeller was foremost among American foundations in support-ing aid and reconstruction activities in Europe during and after World War I.[12] In October 1914 a Rockefeller War Relief Committee was founded and undertook a European study tour that resulted in funding for Hoover's Commission for Relief in Belgium, considerable funds dedicated to expanding the work of the American Red Cross, and the launching of the aforementioned 1915 medical mission to Serbia under the direction of Harvard Medical School professor Dr. Richard P. Strong.[13] Foundation philanthropy played a notable role in American reconstruction projects in Europe: through foundation-managed ini-tiatives as well as grants to existing organizations (e.g., the YMCA, YWCA, and the American Red Cross) and to ad hoc relief organiza-tions (e.g., the Commission for Relief in Belgium, the Serbian Child Welfare Association, among hundreds of others). However, a much larger portion of the American relief work appears to have been funded by individual contributions.

The aid that Americans could offer their European counterparts in the late 1910s and early 1920s also provided an opportunity for the export of the American model of philanthropy in and of itself. Yet, as Merle Curti has pointed out, while philanthropy has been both index and agent of a distinctively "American character," U.S. traditions of charitable giving were initially drawn from Europe and have long borne the imprint of Elizabethan poor laws and empha-sis on providing for the "deserving poor."[14] Recent scholarship on Europe-U.S. "Atlantic crossings" has revealed continuing patterns of mutual interaction,[15] all of which serves as a useful reminder that charitable good works were also a venerable European tradi-tion.[16] However, when, for example, Americans such as Rushton Fairclough, the Red Cross Commissioner to Montenegro, expressed cynicism about Europeans, he drew on a normative vision that linked charity to private initiative and civic volunteerism. This vision—as will be seen below—was understood at the time as the "American" way of doing things. Fairclough, an erstwhile professor of classics at Stanford, in fact laid his critique on Romanians that he observed in 1919 at a horse race in Bucharest, writing,

> these rich people would spend their money freely in amusements, but they would not lift a finger to relieve the poverty of their fellow countrymen. Perhaps, I thought, they are laughing in their sleeves at the generous Americans who have come over here, with their practical philanthropy. How good, I thought, it would be to get back to Montenegro, where everyone is poor, everybody is in rags, and where we know that American money is well spent.[17]

Anxiety not to appear as Lady Bountiful but to develop practical philanthropy that would meet immediate needs as well as allow the planning of a better future pervades the American reconstruction literature. Across the different American reconstruction projects examined in this chapter, one regularly finds the precept that American aid should enable Europeans to become more active in the service of their communities. In one form or another, reconstruction projects helped to transmit a particular vision of the role that charitable works and civic initiative ought to play in civil society. And, as we see in other chapters of this volume, American models spread partly because they were ostensibly offered out of charitable concern and in a manner that professed to be noncoercive (not just in Europe, but globally).

Some of the ways that American reconstruction projects defined problems and solutions and normalized behaviors and subjectivities are nicely illustrated in a letter from an anonymous Frenchwoman published in 1919 in the U.S. social work journal *The Survey*. With armed conflict over, American charities were pressed to make the case that assistance continued to be needed in Europe and thus regularly briefed the American public on the extent of the devastation.[18] In arguing for the continuing need for American aid, the author felt the need to address the possible perception that the French might be unworthy of further assistance. She acknowledged that French initiative and engagement in their own reconstruction did seem lacking, but noted that "Americans are more active in their social work ... because they give less importance to family than the French." The writer accepted the French commitment to family as a fault but explained it as a fact based on obligations to family and children and an education "that makes every one dependent on something or other," adding,

> since the war, many French women and many girls have imitated the Americans and gone out of their families, because it was necessary. All my young friends who are not married work in crèches, dispensaries, canteens, etc.; and all my young friends who are married and mothers do something, some social work, although their situation does not make it always very easy.[19]

While this letter—chosen, of course, by Americans to help make their case—is not without ambiguity, it does demonstrate the potential disruptions and "intrusions" that American reconstruction projects could introduce in Europe. Here, we see differing gender norms cast to some extent in terms of a traditional/modern conflict. In the remainder of this chapter I will examine the various ways that American influence and models were projected in post-conflict reconstruction projects. Social welfare, child welfare, and education quickly emerge as key domains of activity. The chapter proceeds by examining how a civilizing mandate was constructed, then turns to look specifically at the case of norm-making around child welfare and education reform in Serbia.

Market Empire and the Civilizing Mandate

Herbert Hoover was elected U.S. president in 1928 after serving as Secretary of Commerce in both the Harding and Coolidge administrations. He entered office with a record of considerable humanitarian accomplishments, and enjoyed a hero's reputation both in the United States and overseas for his role in coordinating emergency American aid to Europe's distressed civilian population during and after the Great War. A Quaker and former mining engineer who had worked in the American West, Australia, and China, Hoover found himself caught up with the consequences that war had on civilian populations and established the Committee for Relief in Belgium 1914. Throughout the war this remarkable organization conducted its own diplomacy, chartered ships under its own flag, and succeeded in feeding much of occupied Belgium. At the end of the war in 1918 Hoover was named head of the American Relief Administration (ARA) and oversaw health and food initiatives in war-torn Europe. His humanitarian experience was again called upon after WWII when he was asked to establish a U.S. government feeding program in occupied Germany. Hoover's government service was internationally oriented in other dimensions as well: as secretary of commerce he was responsible for doubling the number of commerce offices overseas and for emphasizing U.S. government assistance in the promotion of American exports.[20] Hoover's dual persona as international humanitarian and international businessman stands as emblematic of the overlap between what Victoria de Grazia has usefully termed America's "market empire" and the international humanitarian work that, I am arguing, was undertaken by Americans, at least in the early decades of the twentieth century, as their *mission civilisatrice*.

Classic liberal free trade principles have long provided the warrant for "opening" up foreign markets. Instead of viewing the public sphere as an area that is legitimately controlled and managed by indigenous authorities, Americans have a storied tradition of emphasizing that mercantile commerce can legitimately trump such factors.[21] While the commercial advantages that accrue from such a stance are not to be denied, it also rests on a current of thought (from Hugo Grotius through the Enlightenment) that proposes "exchange" as the most desirable principle to govern international order.[22] And, I would argue, this stance additionally relates to the cultural rules of modern actorhood, that is, the particular, contemporary meanings of human agency and what counts as "agentive" action. Neo-institutionalist sociologists John Meyer and Ronald Jepperson argue that over the last several centuries we have seen the steady relocation of agentic capability away from transcendental or natural authorities into human beings. Rather than looking outside society (to divine design or "forces" of nature) to explain how and why things happen in society, explanations are more and more frequently found within society, within human action. Significantly, Meyer and Jepperson argue that participants in modern society "enact in their identities substantial agency for broad cultural purposes."[23] The uniqueness, in historical terms, of this now increasingly globalized notion of modern actorhood lies in the capability, legitimacy and, at times, even obligation to act in the interests of others.[24] These are the features of "agentive actorhood" that are put into play in many a humanitarian initiative. They also form the cultural rules that allowed American reconstruction workers to act in the interest of Europeans whose options and ability to act had been severely limited by the devastations of the war. While Meyer and Jepperson's work speaks to a generalized modern "cultural system" in fact originating in

Europe, the frequency with which "initiative," "activity," "pioneering spirit," and "*social* work" were attached to Americans in the aftermath of World War I suggests that this was a time when Americans could present themselves as modern actors par excellence.[25] Whether operating in the guise of businessperson or humanitarian, Americans generally took entry into the public sphere as natural right and entitlement.

Sanitation and hygiene make frequent appearance as an arena in which American activity and intervention could uplift Europe. Christopher Endy has discussed the preoccupation with hygiene that appears in the writings of U.S. travelers to Europe around the turn of the twentieth century. He notes that while European backwardness in this regard is bemoaned, the literature also makes mention of the salutary influence that American travelers and their habits have had, particularly in places like France and Italy.[26] Similar accounts can be found in the writings of Americans working on reconstruction projects after World War I. For example, Rushton Fairclough's memoirs describe a project undertaken by one of the members of the American Red Cross Commission to Montenegro where a young American discovered an abandoned Turkish bathhouse, cleaned it out, outfitted it with hot water, and put it back into operation. When the Red Cross Commission was withdrawn in 1921 they turned the operation over to the city of Podgorica, and the whole enterprise was described by Fairclough as "an interesting illustration of American ingenuity and initiative, as well of American cleanliness and comfort."[27] The ways that American sanitation could bring civilization to an "old" Europe were also demonstrated in a report from another Red Cross Commissioner, Ernest P. Bicknell, who wrote of the new Czechoslovak president Tomáš Masaryk discovering that the Austro-Hungarian Hradčany castle in Prague that was now to be his residence did not include a bathroom among its 1,100 rooms. As Bicknell related the story, Masaryk only agreed to live in the castle once an "American bathroom" had been installed.[28]

While there is ample evidence that American firms profited handily from the war and reconstruction projects subsequent to it,[29] my argument is not that American post-conflict aid served as cover for the ulterior motive of expanding overseas markets. The democracy of consumption that underpinned America's Market Empire across the twentieth century (an "imperium with the outlook of an emporium" as de Grazia cleverly puts it) consisted of a set of ideas and practices surrounding what was natural, modern, and best-practice. These were the microphysics of "soft power" that adhered both in commercial and humanitarian activities, and by whose token Masaryk made the right stand. Though they do not issue forth in coordinated fashion from a stable center, the sum of these "rules" and regulative principles have coalesced to afford a position of influence and global preeminence for the United States for much of the twentieth century.[30] Whether in the invention of the calorie,[31] the teaching of new farming techniques, or the dissemination of a health textbook, embedded normative principles rendered people legible and governable according to modern forms of governmentality.

Child Welfare, Vocational Education, and Norm-Making in Serbia

Early in the war, aid to Serbia[32] emerged as *cause célèbre* in the United States. The Serbian war experience had all the makings of high drama: extraordinary heroism and tenacity; a thrice-occupied capital; a victimized peasant population; and, displaced refugees dispersed across Europe and North Africa. In late 1915, having been defeated by the combined forces of the Bulgarian, German, and Austro-Hungarian armies, the Serbian army, together with considerable numbers of civilians, retreated through mountainous Montenegro and Albania. Around 60,000 perished along this route; the 150,000 Serbs that managed to reach the Mediterranean shore were met and relocated by allied transport ships. Americans were heavily involved in aiding Serbian refugees in diaspora,[33] however, in these early stages, American humanitarian involvement in the country itself centered around the American hospital in Belgrade (under the management of Dr. Edward W. Ryan, who was featured in the *New York Times* as "the American Doctor who Saved Belgrade")[34] and Dr. Richard P. Strong's American Red Cross/ Rockefeller Foundation typhus expedition.[35] Reporting on the latter at the Red Cross annual meeting in January 1916, Strong opined that the most important lesson was that the Serbian epidemic would not have occurred "if the Serbs had been properly prepared."[36] Leaving Serbia properly prepared and properly organized in fact became the key objective of the bulk of the post-conflict reconstruction work.

Serbs played an active role in directing American attention to the plight of Serbia. Notable fundraisers included Helen Lozanić[37] who was appointed the Serbian Red Cross delegate to the United States in 1914 and spoke at hundreds of fundraising events across the country over the next five years, raising money for organizations including the Serbian Distress Fund of Boston and the Serbian Relief Fund of

New York (later to become the Serbian Child Welfare Association).[38] Fundraising posters and accounts such as Lozanić's indicate that a discourse of victimization combined with portrayal of Serbians as peasant peoples framed these appeals for aid. Lozanić, the daughter of one of the University of Belgrade's first rectors and Serbian representative at a 1911 international women's conference in Copenhagen, regularly donned colorfully embroidered peasant garb at fundraising events—as did the wife of Serbian ambassador Slavko M. Grujić, also an active fundraiser on behalf of Serbia.[39] In this second case, the attire is even more remarkable given that Mabel Grujić was an American (née Mabel Dunlop) originally from West Virginia. By the late 1910s Serbia had made some strides toward reflexive modernization,[40] however the economy remained overwhelmingly agriculture and the Serbian need for American assistance was cloaked in imagery of peasant purity and primitiveness. In the light of American modernity, Serbia (like other "new" or "young" countries in East/Central Europe) appeared as an "old Europe" whose simplicity and undeveloped state made it particularly suited for American uplift and rescue.

That American aid to Serbia was framed as a civilizing mission is starkly evident in a 1919 account of the Red Cross agricultural unit that worked in Monastir (present-day Bitola, Macedonia). The piece that appeared in the *Red Cross Magazine* was titled "Taking the West into Monastir"—on the one hand an oblique reference to "Western civilization," but more explicitly it was a reference to the involvement of Americans from the upper Midwest. In this instance the delegation was headed by a Croatian immigrant and Catholic priest, Francis Jager, who was also involved in agricultural education at the University of Minnesota.[41] Their vehicles were reported as the "first friendly advance agents of modernity to travel through the streets of many of these huddled villages"[42]; the article included photographs of destitute Serbian refugees returning home in exposed train boxcars as well as a woman and child in native dress. The American mission offered meals, dug wells, offered "technical schooling" and imported farm equipment (tractors, combines, and sawmills) that was eventually turned over to the Serbian government. The article offered the appraisal that thanks to this initiative, "the peasants know something of American machines now; they have American seed and have seen Americans work." Echoing the mythology of the conquest of the frontier in the American West, the piece also spoke to the introduction of

[an] impulse to fruitfulness in the Monastir wilderness, the transplanting of that pioneer American spirit, and the means of

conquering one more stubborn corner of the earth to make it serve man and bear food for his support and enjoyment.[43]

Americans are positioned in this text as the emissaries of "work," agentic capacity and modernity. Imparting these values, dispositions, and ways of being to Serbs was a consistent theme across American reconstruction projects in the country.

As the agricultural mission Monastir evidences, American projects tended to first focus on emergency needs and then aim for long-term impact. This aid trajectory was evident in other theaters of operation, especially on Red Cross projects. In Poland, by 1920/21, the Red Cross program had shifted from general relief to child health and educational work.[44] And, even though the original instructions given to the 1919–21 Red Cross Commission to Montenegro were that their work was to be purely emergency aid and "not to be prompted by any program of a permanent nature,"[45] leaving a long-term legacy became a pressing concern by the end, with the American Junior Red Cross stepping in to continue school and orphanage projects after the commission withdrew.[46] In Serbia, relief work moved from emergency feeding and triaging the needs of displaced people into child health programs and vocational education projects.

The Serbian Child Welfare Association (SCWA) was the principal American organization active in post-conflict reconstruction work in Serbia. While the American Red Cross sent several commissions to Serbia during and after the war, as noted above, the bulk of its postwar work was farmed out to the SCWA, which itself was an outgrowth of the Serbian Relief Fund. Ralph R. Reeder, who had volunteered for the American Red Cross in France and who served superintendent of the New York Orphan Asylum Society, headed the SCWA mission from 1920 to 1922, the period of its principal activity. In Serbia it was often simply referred to as the *Američka misija* or American Commission. In the United States, support for SCWA work grew out of a study tour that Homer Folks, head of Red Cross Relief work in France during the war, took to Italy and the Balkans in late 1918. Folks' report appeared as a book titled *The Human Costs of the War*. Just as Serbia had attracted American attention in the early stages of the war, it did so again after the armistice on account of the desperate circumstances its people faced. However, now this was framed in terms of a nation newly formed and newly embarked on a modernization quest.[47]

The war had extreme consequences for public health. Considerable numbers of Serbian doctors had not survived, and in setting up health

centers in 1920–21, the SCWA was attempting to remedy the near
absence of medical services in some areas. At the same time, the postwar reconstruction period afforded the opportunity for general public
health, sanitation, and hygiene campaigns that the SCWA undertook
at its centers but also on a broader scale by underwriting an elementary
school health textbook[48] and organizing Child Health Exhibits.

Across Europe, the American reconstruction projects that targeted
child health and child welfare were overtly interested in changing
habits and behaviors. In their bids to set new norms, Americans frequently organized didactic exhibits in the spirit of the World's Fairs
and international expositions that had served as spectacles of modernity and progress since the mid-nineteenth century. A fine example
of this was the April 1918 American Red Cross Child Welfare Exhibit
in Lyons, France.[49] Modeled on the Philadelphia Baby Show of 1911,
the Lyons exposition included educational films,[50] a demonstration
kindergarten with actual students, as well as booths dedicated to the
preparation of milk, dental hygiene, and baby care. The Lyons exhibit
recorded 72,000 visitors in its first nine days of operation, and, because
of this success, was turned into a traveling exposition that later toured
Marseilles, Bordeaux, and Paris. In Serbia, the SCWA organized baby
fairs in numerous cities—events that included lessons, demonstrations,
and prizes given to the best cared for babies.[51] Extravagant, expansive
claims were sometimes attached to health-related work: a small Quaker
mission in Peć, Montenegro (present day Kosovo) optimistically proposed that their hospital and regular home visits to develop "the habit
of proper sanitation and better living" could lead to interethnic peace.
"There could be no stronger tool in the work of replacing international hatred and suspicion by trust and understanding," than when
such educational work was coupled with programs that brought Serbs,
Turks, Albanians, and Montenegrans together and trained them to be
nurses, the Quakers claimed.[52] The SCWA work in this area indexed
more predictable arguments about instructing the "Serbian people in
the right methods of infant and child welfare" in order to save babies
and lower the infant mortality rate.

The SCWA's April 1922 Child Health Exhibit in Belgrade was
attended by over 30,000 people and included the participation of local
organizations.[53] This strategy reflected the SCWA's commitment
to partnership projects, an orientation toward reconstruction work
summed up in their position that "whatever you induce a people to do
for themselves is of infinitely more value than what you do for them."[54]
"Co-operative Reconstruction," as it was called, was a hallmark of

SCWA projects. This cooperation had less to do with gaining entry (or justifying intervention) than with building local institutional capacity and exerting an agenda-setting influence. In a 1924 self-appraisal the SCWA concluded that the net result of their child health campaign had been "to 'put the baby on the map' of a public health program in Serbia."[55] Echoing the overlap between market empire and international humanitarian work, SCWA fundraising pamphlets frequently deployed a commercial idiom and spoke of the need to "sell" the Serbian people on their projects.[56]

The SCWA claimed that "it was a point of view, a sense of social obligation rather than the particular thing done that the reconstruction worker 'put across'"[57]—a philosophy that was quite evident in the work undertaken to repair Serbian school buildings. The SCWA school rehabilitation projects were an outgrowth of the child welfare program when it was found that war orphans, while they might be provided with food and shelter, still frequently lacked educational opportunities due to the wartime destruction of school buildings. With funds provided by an anonymous American donor, the SCWA helped to restore 125 school buildings by providing 50 percent of the building funds and requiring local communities to foot the other half of the bill.[58] The SCWA reported:

> In most cases the Serbian peasant communities went far beyond it, for when once aroused and started on the school building job they carried it much further than the Association's initial fifty-fifty proposition required them to go. The blight of war had stupefied the people and paralyzed public spirit. All that was needed to stir latent energies was an initial push from their American friends.[59]

As noted above, American reconstruction projects were clearly—overtly and consciously, as we see here—in the business of exporting a set of ideal behaviors and dispositions related to agency and actorhood.

Local partnerships also undergirded the SCWA's work to advance vocational education in Serbia. Here, the chief mechanism was a local advisory board, the "Committee on Institutions and Vocational Education," minutes from whose meetings show that it did not merely rubber-stamp but was quite active in shaping SCWA activities.[60] As is shown by the involvement of Serbs, not to mention the industrial schools that had been in operation before the war, it is far from the case that Americans single-handedly introduced vocational education to Serbia. Nonetheless, it is worth noting that Americans advanced and advocated such forms of schooling in their reconstruction work.

Part of the explanation for this lies in a belief in the social and individual benefits of work—something that we find in numerous American aid initiatives in Europe. Early in the war, the Red Cross undertook some experiments in providing employment opportunities for female Belgian refugees in Holland. When provided with sewing machines and instructed in the manufacture of undergarments, a "transformation" occurred: "the opportunity to provide their families and others with warm underwear" revolutionized the discipline problems that had previously plagued the camps. Under the supervision of American pastor John Van Schaick, the experiment was carried into 35 refugee camps, where it was found that sewing and knitting classes had "counteracted the demoralizing influence of refugee life," brought about friendships between Belgian and Dutch women, and, overall, had "both an educational and moral influence."[61] In Switzerland, the American Red Cross set up a trade school for displaced Belgian boys who were known after their benefactor as "Rockefeller children."[62] The practical and moral benefit or replacing idle hands with industrious activity was abundantly apparent in refugee situations and continued to be a strategic objective in post-conflict reconstruction as well.

American sponsorship of vocational education in post-conflict Serbia also reflects the broader pattern of overseas American educational work strongly favoring manual training and vocational education.[63] As Jason Yaremko argues in this volume, and historian Jonathan Zimmerman has also shown, the U.S. international emphasis on agricultural and industrial subjects over academic ones extended well beyond the well-known Tuskegee-in-Africa projects sponsored in the 1920s and 1930s by the Phelps-Stokes Fund. From the end of the nineteenth century, in the Philippines, Haiti, and Puerto Rico, American officials advanced the cause of vocational education. American missionaries in the Middle East, Cuba, and in India also had a long tradition of establishing vocational schools.[64] In the post-conflict setting of Serbia after World War I, vocational education held the promise of individual rescue and social development. An article on Serbia published in *The Survey* in 1919 noted the need for more agricultural education, called for a university extension and rural high school system, and proposed that this was an area in which "the genius of America is most likely to find a congenial sphere of remarkable usefulness."[65]

Through grants and equipment donated by the American Red Cross and the American Relief Administration, the SCWA assisted in the reopening of Serbian trade schools. Tools, yarn, and sewing machines were provided to over 200 trade schools and scholarships were provided for several hundred war orphans. In this work the SCWA (the "American Commission") positioned itself as agent of broad American support for Serbia. In some cases American and Serbian institutions were paired, as was the case of the Valjevo Agricultural School, which was linked with the Boys High School of Brooklyn whose students donated all the needed funds. The Serbian letters of thanks reprinted by the SCWA register an appreciation of this assistance as *American* assistance, including expressions of gratitude, for example, to the "great and noble American nation."[66] Promotional fundraising material produced by the SCWA to appeal to Americans illustrates how American interests were tied to Serbian reconstruction. Progress, haunted by the specter of regress, framed the appeals as we see in figure 4.1. Surrounding the textual argument that "the work MUST go on," is imagery suggesting

Figure 4.1 Serbian Child Welfare Association Promotional Pamphlet. (Serbian Child Welfare Association, undated pamphlet [1921?], "The Work Must Go On." By courtesy of the Department of Special Collections, Memorial Library, University of Wisconsin-Madison.)

that the advancement of the Serbian peasantry will come through educational work, namely through the "American Industrial School" of
Čačak whose monumental edifice tops the image. When the SCWA
published images of children at work inside the trade schools and the
"domestic science schools" that were also supported,[67] they included
captions such as "the great ambition of Serbians today, from youngest to oldest, is for practical training."[68] Such grand claims should
be approached with skepticism, particularly as Zimmerman has clearly
demonstrated that the American desire to spread vocational education
across the globe was regularly confronted with strong local opposition
and frequent preference for "academic" schooling tracks.[69] It would
certainly be a mistake to assume that these projects and messages were
simply passively received in Serbia. The "American" models of agentic
action, programs of child welfare, vocational education schools, and the
dispositions and habits that went with them took some form in Serbia,
yet this was not without resemanticization, recontextualization, and
perhaps even resistance.[70]

Conclusion: American Modernity Received?

Their ideas were too grand: Serbia was not ready for them. They
came over in 1919 and 1920 with high hopes and plans for setting
up a model Child Welfare Service all over the country. While they
were on the Atlantic they decided where all the centres would be.
They drew red and blue circles on their maps and plotted the
whole thing out. They intended to have ten different centres with
outposts dependent on them, clearing-houses for abandoned children, model orphanages, Infant Wellness Centres, even Vocational
Guidance Clinics and Homes for the deaf and dumb—they forgot
nothing. They were a disciplined, well-trained body, but when
their scheme collapsed they did not know how to take up something else: they were not adaptable.... On the whole they were
disappointed. The trouble was they wanted to do things too much
as Americans, and the Serbs were bursting with energy and national
pride and did not want anything imposed from without.[71]

This appraisal was offered by Margaret McFie, a Scottish woman
who worked on relief projects in Serbia during and after the war.
Predictably, the SCWA accounts painted a rosier picture.[72] Ralph

Reeder, upon returning to Serbia in 1923 observed that a number of the SCWA projects, including the domestic science schools, the Nurses Training School the Americans had founded in Belgrade, as well as a good number of local health centers all continued under local management. When former Red Cross commissioner to Montenegro, Rushton Fairclough, returned in 1931 he found that his 1920 "comprehensive plan" for medical and educational initiatives had been partially put into effect. He found that some substitutions had been made (e.g., a hospital opened in Danilovgrad in place of Kolashin) but he generally considered all to be "operating in accordance with a well-conceived plan in which the Rockefeller Foundation played an important part."[73] Yet, despite disagreement on whether the schemes ultimately collapsed or succeeded, the American accounts do correspond with McFie's impression that on an institutional level the Americans aimed not for stopgap assistance but for extensive restructuring.

Across Europe, American institutional restructuring aimed to systematize social welfare provision. In 1918, before the war's end, the *Red Cross Magazine* reported that "French soil is showing itself wholly congenial" to the introduction of a Social Service Exchange (a centralized information repository that allowed coordination among relief agencies and eliminated duplication, at the time a social work best-practice in American cities).[74] In a study of the influence of American women who worked on reconstruction projects in the Soissons area of France, Evelyne Diebolt and Nicole Fouché argue that the Americans had significant effects in the area of public health, notably through the creation of charitable associations and public-private partnerships.[75] The "social survey" was also widely used in American reconstruction work, in Italy[76] and perhaps most notably in Czechoslovakia where Masaryk's daughter, Alice Garrigue Masaryk, who had earlier spent a year and a half living at Hull House in Chicago (1904–5), made arrangements with Mary McDowell for an American trained social worker to undertake a social survey of Prague and to assist in developing the field of social work in the country.[77] In Serbia, beyond what is recounted above in reference to the SCWA work and the mission Fairclough directed, the clearest evidence of a lasting institutional legacy following from American post-conflict reconstruction can be seen in the Yugoslav Ministry of Health, specifically in the work of the head of the Department of Hygiene and Social Medicine, Andrije Štampar, who had extremely close ties with the Rockefeller Foundation.[78] In Austria, the postwar American Red Cross commission was purported to have accomplished "a complete renovation and reorganization of the already established child welfare activities," yet Austria also provides one of the few instances where American relief workers acknowledged European proficiency: in public health "the Americans had little or nothing to teach Austria in the way of organization or in methods of treatment."[79]

Such expressions of humility are not to be found in the American literature on Serbia and its reconstruction needs. Though the technical expertise of some Serbian doctors and officials was conceded,[80] the SCWA lamented the lack of initiative and industriousness in their Serbian counterparts. Ralph Reeder's internal bulletins to SCWA personnel in Serbia highlight this, as when he exhorted his employees that for Americans "there ain't no such word as *ne mozhe* [it's not possible]." He also advised that Americans overseas should avoid "infection from the *sutra* [tomorrow] disease, for it's like malaria, hard to eradicate from the blood."[81] Actorhood capacity was, of course, one of the very things that Americans sought to advance in Serbia. The school building program and the community "energy" that American aid released from its previous "paralysis" would be one area where the SCWA made the claim that they were successful in "putting across" notions of social obligation. However, despite the claims that its reconstruction program was "a program of real COMMUNITY BUILDING, the consummation of all the organized efforts of the various social, sanitary and cultural groups of the government and the country at large,"[82] on the surface, the SCWA does not seem to have markedly rewired local cultural patterns of agentic action.

Across the 1920s and 1930s, as I have argued elsewhere,[83] Yugoslavia tended to look toward other Slavic countries to find exemplars of modernity. Notions of "Slavic" cultural affinity and coevality made Czechoslovakia (and Poland to a lesser extent) attractive referent societies for Yugoslav modernization projects. Of course, cultural projections of America, "American" behaviors and "American" attitudes did not vanish from the picture. Comments from a Yugoslav education professor who traveled to the Czechoslovak city of Zlín in 1933 provide one good—if slightly enigmatic—example of how "American modernity" could be recontextualized. Zlín was home to the Bata shoe company; it was a factory town with "modern" schools, social welfare provisions and civic institutions. Zagreb Professor of Pedagogy Salih Ljubunčić observed, "Zlín is a piece of America in Czechoslovakia—a pure Slavic America, not the self-estranged Anglo-Saxon America, but a real America."[84] Ljubunčić's semantic recoding appropriated an authentic modernity for Central/ Eastern Europe. And, while we do not know with certainty the extent to which American postwar reconstruction shaped this statement, it does suggest that the "Americanization," which was brought to Europe through these initiatives, was reworked (hybridized and creolized) according to local circumstances. It is not difficult to posit that such a conclusion is likely to hold for most "civilizing missions." The chief objective of this chapter has been to establish what values, behaviors, and principles were embedded in the work of the "third American army." In short, philanthropically-minded American relief workers attended to health and nutritional needs but also attempted to educate Europeans in self-reliance and social responsibility. In the work done on sanitation, hygiene, vocational education, and child welfare provision, domains of expertise and areas of American superiority were carved out. American reconstruction work in the aftermath of World War I had consequences for the ways that Europeans reflected on themselves, envisioned the future and attempted to realize it. At the same time, this had consequences for America's self-image. For, having brought freedom, uplift and civilization to Europe, Americans could increasingly see themselves as legitimated in projecting norms and best-practices around the globe.

Notes

1. Serbian Child Welfare Association of America, *Co-operative Reconstruction: A Report of the Work Accomplished in Serbia* (New York: Serbian Child Welfare Association of America, 1924), 98. Though published without attribution, this text is most certainly the work of Ralph R. Reeder, the organization's overseas commissioner who was based in Serbia (then part of the Kingdom of Serbs, Croats and Slovenes) in the early 1920s. An earlier draft of this text dated February 21, 1921, can be found in a folder containing Reeder's correspondence in the Edward R. Johnstone papers, box 1, (MC 538) in the Rutgers University Special Collections Library. Hereafter cited as Johnstone papers, Rutgers Special Collections.
2. Charles H. Hopkins, *History of the Y.M.C.A. in North America* (New York: YMCA, 1951), 485–502. An in-house estimate holds that American Red Cross field hospitals treated more than a third of American battle casualties. See American National Red Cross. Nursing Service et al., *History of American Red Cross Nursing* (New York: The Macmillan Company, 1922), 305. More than the YMCA, the YWCA appears to have engaged in civilian relief projects alongside the aid rendered as part of the U.S. war effort. See Nancy Boyd, *Emissaries: The Overseas Work of the American YWCA 1895–1970* (New York: Woman's Press, 1986), 72–76.
3. The best general treatment remains Merle Eugene Curti, *American Philanthropy Abroad: A History* (New Brunswick, NJ: Rutgers University Press, 1963), 224–300. For American volunteer activity during the war years see also, Edwin W. Morse, *The Vanguard of American Volunteers in the Fighting Lines and in Humanitarian Service August, 1914–April, 1917* (New York: C. Scribner's Sons, 1918), 95–202.
4. For a good general account see J. P. Greene, *The Intellectual Construction of America: Exceptionalism and Identity from 1492 to 1800* (Chapel Hill, NC: University of North Carolina Press, 1993).

5. Victoria de Grazia, *Irresistible Empire: America's Advance through Twentieth-Century Europe* (Cambridge, MA: Belknap Press of Harvard University Press, 2005), 4–9.
6. Peter Wagner, "The Resistance the Modernity Constantly Provokes: Europe, America and Social Theory," *Thesis Eleven* 58 (1999): 45–46. To be sure, the Great War was only one among numerous events or moments that caused Europeans to reflect on American modernity (something the Paris World's Fair of 1900 clearly did). However, it can be argued, as Mary Nolan does in her study of the modernization of German business and industry, that by the 1920s Europeans were approaching America less with bemused skepticism than in search of answers. See Mary Nolan, *Visions of Modernity: American Business and the Modernization of Germany* (Oxford: Oxford University Press, 1994), 17, 23–26. Note that my argument here pertains to European self-definition. Many Americans viewed European travel as museum-going and ascribed "old" status to Europe long before this. See James Buzard, *The Beaten Track: European Tourism, Literature, and the Ways to Culture 1800–1918* (Oxford: Clarendon Press, 1993), 71–81.
7. Volker R. Berghahn, "Philanthropy and Diplomacy in the 'American Century,' " *Diplomatic History* 23, no. 3 (1999).
8. David C. Engerman, *Staging Growth: Modernization, Development, and the Global Cold War, Culture, Politics, and the Cold War* (Amherst, MA: University of Massachusetts Press, 2003); Nils Gilman, *Mandarins of the Future: Modernization Theory in Cold War America* (Baltimore, MD: Johns Hopkins University Press, 2004).
9. John Farley, *To Cast Out Disease: A History of the International Health Division of the Rockefeller Foundation (1913–1951)* (Oxford; New York: Oxford University Press, 2004).
10. Ellen Condliffe Lagemann, *The Politics of Knowledge: The Carnegie Corporation, Philanthropy, and Public Policy*, 1st ed. (Middletown, CT: Wesleyan University Press, 1989). The significance of the involvement of American foundations in knowledge production is also clearly demonstrated in studies of Rockefeller's interwar funding of child development research, specifically for the ways that this worked to advance mental hygiene perspectives in psychology and pedagogy. See Brian Low, "The Hand that Rocked the Cradle: A Critical Analysis of Rockefeller Philanthropic Funding, 1920–1960," *Historical Studies in Education/ Revue d'histoire de l'éducation* 16, no. 1 (2004).
11. Lagemann, *The Politics of Knowledge*, 67.
12. All told, the Rockefeller Foundation spent more than $22 million on war-relief activities. In this same period Americans contributed $34 million to Hoover's Committee on Belgian Relief, $63 million was raised to aid Jews in Europe, and the Red Cross garnered approximately $400 million in donations. See Curti, *American Philanthropy Abroad*, 235, 44, 48; Raymond Blaine Fosdick, *The Story of the Rockefeller Foundation*, 1st ed. (New York: Harper, 1952), 28.
13. Ernest Percy Bicknell, *In War's Wake, 1914–1915: The Rockefeller Foundation and the American Red Cross Join in Civilian Relief* (Washington, DC: American National Red Cross, 1936).
14. Merle Curti, "American Philanthropy and the National Character," *American Quarterly* 10, no. 4 (1958): 424–429.
15. Daniel T. Rodgers, *Atlantic Crossings: Social Politics in a Progressive Age* (Cambridge, MA: Harvard University Press, 1998). See also William J. Reese, "After Bread, Education: Nutrition and Urban School Children, 1890–1920," *Teachers College Record* 81 (1980).
16. Beyond the long-standing tradition of aristocratic endowments, we also find the involvement of European capitalist businessmen in philanthropy and public improvements. See John Brewer, "Selling the American Way/Book Review of de Grazia 'Irresistible Empire: America's Advance through Twentieth-Century Europe,' " *New York Review of Books* 53, no. 19 (2006): 60. Nonetheless, for a discussion of contrasts in English and American philanthropy see John Hamer, "English and American Giving: Past and Future Imaginings," *History and Anthropology* vol. 18, no. 4 (2007).

17. Rushton H. Fairclough, *Warming Both Hands: The Autobiography of Henry Rushton Fairclough, Including His Experiences under the American Red Cross in Switzerland and Montenegro* (Stanford, CA: Stanford University Press, 1941), 353.
18. See, for example, Jane Addams and Alice Hamilton, "After the Lean Years: Impressions of Food Conditions in Germany When Peace was Signed," *The Survey* 42, no. 23 (1919); Homer Folks, *The Human Costs of the War* (New York: Harper & Brothers Publishers, 1920).
19. "America Overseas: Our Efforts in France," *The Survey* 42, no. 1 (1919): 48.
20. Ulf Jonas Bjork, "The U.S. Commerce Department Aids Hollywood Exports, 1921–1933," *Historian* 62, no. 3 (2000); Frank Costigliola, *Awkward Dominion: American Political, Economic, and Cultural Relations with Europe, 1919–1933* (Ithaca, NY: Cornell University Press, 1984); Emily S. Rosenberg and Eric Foner, *Spreading the American Dream: American Economic and Cultural Expansion, 1890–1945*, 1st ed., American Century Series (New York: Hill and Wang, 1982). See also Joseph Brandes, "Product Diplomacy: Herbert Hoover's Anti-Monopoly Campaign at Home and Abroad," in *Herbert Hoover as Secretary of Commerce*, ed. Ellis W. Hawley (Iowa City, IA: University of Iowa Press, 1981).
21. Strongly protectionist policies in the United States itself notwithstanding, de Grazia, *Irresistible Empire*, 7.
22. For a discussion of the premium Benjamin Franklin placed on "exchange" see Jim Egan, "Turning Identity Upside Down: Benjamin Franklin's Antipodean Cosmopolitanism," in *Messy Beginnings: Postcoloniality and Early American Studies*, ed. Malini Johar Schueller and Edward Watts (New Brunswick, NJ: Rutgers University Press, 2003).
23. John W. Meyer and Ronald L. Jepperson, "The 'Actors' of Modern Society: The Cultural Construction of Social Agency," *Sociological Theory* 18, no. 1 (2000): 101.
24. Actorhood on behalf of others could extend from representing the interests of those with a perceived limited agency (e.g., the "poor" or until recently the disabled), as well the interests of nonactors (e.g., "the environment," spotted owls).
25. The "Americanization" of agentic actorhood was not, of course, confined to years just after World War I as de Grazia's account of the spread of Rotary clubs throughout Europe would suggest.
26. Christopher Endy, "Travel and World Power: Americans in Europe, 1890–1917," *Diplomatic History* 22, no. 4 (1998): 582–583.
27. Fairclough, *Warming Both Hands*, 365.
28. Ernest Percy Bicknell and Grace Vawter Bicknell, *With the Red Cross in Europe, 1917–1922* (Washington, DC: American National Red Cross, 1938), 366–367. I have been unable to confirm or refute Bicknell's anecdote. However, in the interwar period the Prague castle was in fact renovated on Masaryk's orders to become a "democratic castle" and included a presidential apartment with bathroom. See Jana Horneková, "Plečnik's construction of T. G. Masaryk's apartment," in *Josip Plečnik—An Architect of Prague Castle*, ed. Zeděnek Lukeš and Damjan Prelovšek (Prague: Prague Castle Administration, 1996).
29. For an analysis of this war profiteering in relation to recent conflicts see Saltman, this volume.
30. In their work on "empire," Antonio Negri and Michael Hardt are quick to argue that the United States is not at the center of the contemporary imperialist project that they maintain accurately describes the globalization processes of the early twenty-first century. I am using the notion of empire differently—not to speak to a general political and social condition, but rather to think about the specific circulations of power where geographic relations, and certainly their cultural imaginings, are significant. See Michael Hardt and Antonio Negri, *Empire* (Cambridge, MA: Harvard University Press, 2000).
31. Nick Cullather, "The Foreign Policy of the Calorie," *American Historical Review* 112, no. 2 (2007).

96   *Noah W. Sobe*

32. The Kingdom of Serbia had formally existed since 1882, though certain degrees of independence from the Ottomans were achieved earlier in the nineteenth century. And, even as "the first" Yugoslavia came into existence in December 1918 (and was officially known until 1929 as the Kingdom of the Serbs, Croats and Slovenes), well into the 1920s American sources sometimes speak of "Serbia" as shorthand for the entire country. While this tendency is evident in SCWA documents, their work was principally focused on the region known as Serbia, which is the principal focus of the discussion that follows here.
33. In December, 1916 editors of the *Red Cross Magazine* remarked that "One of the most useful—and most appreciated—works of American philanthropy in the European War has been the relief of the Serbians who retreated through Albania and found refuge on Corfu and neighboring islands in the Adriatic sea." Ellwood Hendrick, "The Serbians," *Red Cross Magazine* 11, no. 12 (1916): 415.
34. "American Doctor Saved Belgrade," *New York Times*, January 31, 1915.
35. The American Women's Hospitals Service was also active in the southern part of the country (now Kosovo and Macedonia), with some of their projects transferring to the Serbian Child Welfare Association in 1920. See Esther Pohl Lovejoy, *Certain Samaritans*, New, rev and reset. ed. (New York: Macmillan Company, 1933).
36. "An Interesting Afternoon Session," *Red Cross Magazine* 11, no. 1 (1916): 21.
37. Lozanič (whose surname is frequently rendered in English-language sources as Losanitch) later married John Frothingham a Red Cross official stationed in Serbia. In the 1920s and early 1930s the two supported orphanages in Vranje and Kamenitza.
38. Helen Losanitch Frothingham and Matilda Spence Rowland, *Mission for Serbia: Letters from America and Canada, 1915–1920.* (New York: Walker, 1970).
39. See the photograph appearing in "American Doctor Saved Belgrade." Mabel Grujić(Grouitch) solicited aid to Serbia while her husband was Serbian ambassador to London. She both mobilized British volunteers and also crossed the Atlantic to raise funds in the United States—initially, as early as 1912 when Serbia was embroiled in the First Balkan War. See Monica Krippner, "The Work of British Medical Women in Serbia during and after the First World War," in *Black Lambs and Grey Falcons: Women Travellers in the Balkans*, ed. John B. Allcock and Antonia Young (New York: Berghahn, 2000), 73; "Mme. Grouitch Here for A Servian Fund," *The New York Times*, October 28, 1912.
40. Arsen Djurović, *Kosmološko Traganje za Novom Školom: Modernizacioni izazovi u sistemu srednjoškolskog obrazovanja u Beogradu 1880–1905* (Beograd: Beogradski Izdavačko-Grafički Zavod, 1999).
41. Jager's mission clashed with the Red Cross work done by Dr. Edward W. Ryan, with each apparently accusing the other of improprieties. See Bicknell and Bicknell, *With the Red Cross in Europe*, 136, 73–77. In the case of the Balkans, there do not seem to have been considerable numbers of American immigrants returning to the homeland to assist in reconstruction projects. One of the most renowned instances of this involved 75 Polish-American women trained in a YWCA program in New York City and then sent to work alongside the American Relief Administration in Poland, see Robert Szymczak, "An Act of Devotion: The Polish Grey Samaritans and the American Relief Effort in Poland, 1919–1921," *Polish American Studies* 43, no. 1 (1986). The "Old Country Schools" operated in this period by the YWCA to train female American immigrants for social work service in their countries of origin do not seem to have included significant Yugoslav participation. See Nancy Gentile Ford, "The Old Country Service School: Gender, Class, and Identity and the YWCA's Training of Immigrant Women in International Social Welfare Philosophy, 1919," *Peace & Change* 23, no. 4 (1998).
42. Lyman Bryson, "Taking the West into Monastir," *Red Cross Magazine* 14, no. 7 (1919): 69.
43. Ibid., 72.
44. Bicknell and Bicknell, *With the Red Cross in Europe*, 351.
45. Fairclough, *Warming Both Hands*, 324.

46. Bicknell and Bicknell, *With the Red Cross in Europe*, 487.

47. Homer Folks, who served as a SCWA board member, also suggested in his May 1920 book that Serbs bore a curious likeness to Americans, a perception that may have furthered the idea that Serbia was fruitful ground for American influence. "They look like Americans, talk like Americans, and seem to think like Americans," he noted. Nonetheless, this also may have been no more than hackneyed national stereotyping, for Folks additionally appraised Serbs as "like the Japanese in their desire to learn the best quickly from other peoples; like the French in scrupulous politeness and deference; like the Italians in the warmth of their welcome and the frank expression of their sentiments; like the English in their dogged resistance; and like the Yankees in their rugged individualism." Folks, *The Human Costs of the War*, 16, 97.

48. Space limitations preclude a full discussion of the politics that surrounded this textbook, however archival records connected with it are revealing in that they provide evidence of a local acknowledgment of America's advancedness when it came to health, sanitation, and hygiene, as well as evidence of interregional tensions—such as on the question of whether the book would come out in Cyrillic or Roman script. See Archives of Yugoslavia (Belgrade), Fund 66, Box 2305, Folder 2176.

49. William Palmer Lucas, "For the Children of France," *Red Cross Magazine* 13, no. 8 (1918): 62–68; Mary Ross, "An American Health Exhibit in France," *The Survey* 40, no. 16 (1918): 449–450.

50. Traveling "motion picture" exhibits were a common public health reconstruction strategy, seen also in Serbia and Czechoslovakia. See Bicknell and Bicknell, *With the Red Cross in Europe*, 476.

51. Prize-giving was a staple of the "better babies" health campaigns in the United States and was widely used in American reconstruction work across Europe. See, for example, "Visiting Carpathian Villages," *International Service: Bulletin of the Society of Friends' Relief Missions in Europe*, no. 4 (1920).

52. "Notice," *International Service: Bulletin of the Society of Friends' Relief Missions in Europe*, no. 3 (1920).

53. Serbian Child Welfare Association of America, *Co-operative Reconstruction*, 25–28. See also, Bicknell and Bicknell, *With the Red Cross in Europe*, 498.

54. Emphasis in original. Serbian Child Welfare Association of America, *Co-operative Reconstruction*, 108.

55. Ibid., 26.

56. See Serbian Child Welfare Association, undated, "Should the Work Go On?" and "The Work Must Go On" in *Pamphlets on Serbian Relief* (F0884 SE6 Cutter), Special Collections, University of Wisconsin Madison.

57. Serbian Child Welfare Association of America, *Co-operative Reconstruction*, 94.

58. This program bears remarkable similarities to the Rosenwald Fund's (1914 through mid-1930s) projects to rehabilitate schools for blacks in the American South. Though Julius Rosenwald was active in postwar reconstruction with the American Red Cross and the Joint Distribution Committee organized to aid Jews in Central/Eastern Europe, I have not been able to identify any direct links between Rosenwald and the SCWA. It is worth mentioning that the scheme of 50/50, foundation/local matching funds was also a defining characteristic of Carnegie's Library building program, which operated from 1883 into the 1920s. For a discussion of the Rosenwald initiative see James D. Anderson, *The Education of Blacks in the South, 1860–1935* (Chapel Hill, NC: University of North Carolina Press, 1988), 148–185.

59. Serbian Child Welfare Association of America, *Co-operative Reconstruction*, 32.

60. See, for example, "Meeting of the Committee on Institutions and Vocational Education," dated January 26, 1921 in Box 1, Johnstone papers, Rutgers Special Collections.

61. Bicknell, *In War's Wake, 1914–1915*, 73–76.

62. Fairclough, *Warming Both Hands*, 297.
63. The late 1910s also represent one of the high points in American educators' infatuation with vocational education on the home front, something we see in the passage of the Smith-Hughes National Vocational Education Act of 1917 and the release of the Cardinal Principles report in 1918. See the discussion in, Herbert M. Kliebard, *Schooled to Work: Vocationalism and the American Curriculum, 1876–1946* (New York: Teachers College Press, 1999).
64. Jonathan Zimmerman, *Innocents Abroad: American Teachers in the American Century* (Cambridge, MA: Harvard University Press, 2006), 51–56.
65. "America Overseas: For a Serbian Cornell," *The Survey* 41, no. 18 (1919): 632.
66. Serbian Child Welfare Association of America, *Co-operative Reconstruction*, 39.
67. SCWA projects to establish and support domestic science schools grew out of the organization's health program and the Americans' dismay at Serbians' "utter ignorance of a better and more wholesome way of living." Ibid., 72–74.
68. Ibid., facing p. 40.
69. Zimmerman, *Innocents Abroad*, 58–61.
70. See the discussion of the problem of "reception" in Rob Kroes, "American Empire and Cultural Imperialism: A View from the Receiving End," *Diplomatic History* 23, no. 3 (1999).
71. Margaret McFie quoted in Francesca M. Wilson, *In the Margins of Chaos: Recollections of Relief Work in and between Three Wars* (New York: Macmillan Company, 1945), 112.
72. I find fault with Merle Curti's conclusion that "many Americans working in Serbia were ... disappointed with what they accomplished." Curti, *American Philanthropy Abroad*, 263. The evidence I have examined here suggests that, while American designs weren't perfectly realized, many successes were trumpeted with considerably self-satisfaction.
73. Fairclough, *Warming Both Hands*, 412.
74. Known in France as the *Fichier Central* and in the United States also as a Charities Clearing House and Confidential Exchange "America Overseas: The Confidential Exchange in Paris," *The Survey* 41, no. 5 (1918).
75. Evelyne Diebolt and Nicole Fouché, "1917–1923, les Américaines en Soissonais: leur influence sur la France," *Revue Française d'Études Américaines*, no. 59 (1994).
76. Richard A. Bolt, "America Overseas: For the Children of Italy," *Red Cross Magazine* 42, no. 14 (1919): 544–546.
77. Julia Lathrop and Grace Abbot were also involved in creating a new childcare bureau in Czechoslovakia, making this a fascinating instance of women's international political activism that merits further study. See "Czechoslovakia Seeks to be Little America," *New York Times*, June 19, 1921; Alice Garrigue Masaryk and Ruth Crawford Mitchell, *Alice Garrigue Masaryk 1879–1966: Her Life as Recorded in Her Own Words and by Her Friends* (Pittsburgh, PA: University of Pittsburgh, 1980); Howard Wilson, *Mary McDowell: Neighbor* (Chicago, IL: University of Chicago Press, 1928), 220.
78. See Linda Killen, "The Rockefeller-Foundation in the 1st Yugoslavia," *East European Quarterly* 24, no. 3 (1990); Patrick Zylberman, "Fewer Parallels than Antitheses: René Sand and Andrija Stampar on Social Medicine, 1919–1955," *Social History of Medicine* 17, no. 1 (2004).
79. Bicknell and Bicknell, *With the Red Cross in Europe*, 473, 71.
80. On his 1918–19 study tour Homer Folks fell ill in Belgrade and reported high satisfaction with the medical care he received. Folks, *The Human Costs of the War*, 18–19.
81. Commission Personnel Letter No. 4 [February 23, 1921], Box 1 Johnstone papers, Rutgers Special Collections.
82. Emphasis in original. Serbian Child Welfare Association of America, *Co-operative Reconstruction*, 112.

83. Noah W. Sobe, "Balkanizing John Dewey," in *Modernities, Inventing The Modern Self and Education: The Traveling of Pragmatism and John Dewey*, ed. Thomas S. Popkewitz (New York: Palgrave, 2005); Noah W. Sobe, "Slavic Emotion and Vernacular Cosmopolitanism: Yugoslav Travels to Czechoslovakia in the 1920s and 1930s," in *Turizm: The Russian and East European Tourist under Capitalism and Socialism*, ed. Anne E. Gorsuch and Diane P. Koenker (Ithaca, NY: Cornell University Press, 2006).

84. Salih Ljubunčić, *Školstvo i prosvjeta u Čehoslovačkoj: s osobitim obzirom na pedagošku i školsku reformu*, ed. Salih Ljubunčić, *Biblioteka "Škole Rada"* (Zagreb: Naklada A. Brusina Naslj. V. i M. Steiner, 1934), 9. See my discussion of this study-tour in, Noah W. Sobe, "Cultivating a 'Slavic Modern': Yugoslav Beekeeping, Schooling and Travel in the 1920s and 1930s," *Paedagogica Historica* 41, no. 1+2 (2005).

Bibliography

American National Red Cross. Nursing Service, Lavinia L. Dock, Sarah Elizabeth Pickett, and Clara D. Noyes. *History of American Red Cross Nursing*. New York: The Macmillan Company, 1922.

Anderson, James D. *The Education of Blacks in the South, 1860–1935*. Chapel Hill, NC: University of North Carolina Press, 1988.

Berghahn, Volker R. "Philanthropy and Diplomacy in the 'American Century.'" *Diplomatic History* 23, no. 3 (1999): 393–419.

Bicknell, Ernest Percy. *In War's Wake, 1914–1915: The Rockefeller Foundation and the American Red Cross Join in Civilian Relief*. Washington, DC: American National Red Cross, 1936.

Bicknell, Ernest Percy and Grace Vawter Bicknell. *With the Red Cross in Europe, 1917–1922*. Washington, DC: American National Red Cross, 1938.

Bjork, Ulf Jonas. "The U.S. Commerce Department Aids Hollywood Exports, 1921–1933." *Historian* 62, no. 3 (2000): 575–587.

Boyd, Nancy. *Emissaries: The Overseas Work of the American YWCA 1895–1970*. New York: Woman's Press, 1986.

Brandes, Joseph. "Product Diplomacy: Herbert Hoover's Anti-Monopoly Campaign at Home and Abroad," in *Herbert Hoover as Secretary of Commerce*, ed. Ellis W. Hawley, 185–216. Iowa City, IA: University of Iowa Press, 1981.

Brewer, John. "Selling the American Way/Book Review of de Grazia'Irresistible Empire: America's Advance through Twentieth-Century Europe.'" *New York Review of Books* 53, no. 19 (2006): 58–61.

Buzard, James. *The Beaten Track: European Tourism, Literature, and the Ways to Culture 1800–1918*. Oxford: Clarendon Press, 1993.

Costigliola, Frank. *Awkward Dominion: American Political, Economic, and Cultural Relations with Europe, 1919–1933*. Ithaca, NY: Cornell University Press, 1984.

Cullather, Nick. "The Foreign Policy of the Calorie." *American Historical Review* 112, no. 2 (2007): 337–364.

Curti, Merle. "American Philanthropy and the National Character." *American Quarterly* 10, no. 4 (1958): 420–437.

Curti, Merle Eugene. *American Philanthropy Abroad: A History*. New Brunswick, NJ: Rutgers University Press, 1963.

de Grazia, Victoria. *Irresistible Empire: America's Advance through Twentieth-Century Europe*. Cambridge, MA: Belknap Press of Harvard University Press, 2005.

Diebolt, Evelyne, and Nicole Fouché. "1917–1923, les Américaines en Soissonais: leur influence sur la France." *Revue Française d'Études Américaines*, no. 59 (1994): 45–63.

Djurović, Arsen. *Kosmološko Traganje za Novom Školom: Modernizacioni izazovi u sistemu srednjoškolskog obrazovanja u Beogradu 1880–1905*. Beograd: Beogradski Izdavačko-Grafički Zavod, 1999.

Egan, Jim. "Turning Identity Upside Down: Benjamin Franklin's Antipodean Cosmopolitanism," in *Messy Beginnings: Postcoloniality and Early American Studies*, ed. Malini Johar Schueller and Edward Watts, 203–222. New Brunswick, NJ: Rutgers University Press, 2003.

Endy, Christopher. "Travel and World Power: Americans in Europe, 1890–1917." *Diplomatic History* 22, no. 4 (1998): 565–594.

Engerman, David C. *Staging Growth: Modernization, Development, and the Global Cold War, Culture, Politics, and the Cold War*. Amherst, MA: University of Massachusetts Press, 2003.

Fairclough, Rushton H. *Warming Both Hands: The Autobiography of Henry Rushton Fairclough, Including His Experiences under the American Red Cross in Switzerland and Montenegro*. Stanford, CA: Stanford University Press, 1941.

Farley, John. *To Cast Out Disease: A History of the International Health Division of the Rockefeller Foundation (1913–1951)*. Oxford; New York: Oxford University Press, 2004.

Fosdick, Raymond Blaine. *The Story of the Rockefeller Foundation*. 1st ed. New York: Harper, 1952.

Frothingham, Helen Losanitch and Matilda Spence Rowland. *Mission for Serbia; Letters from America and Canada, 1915–1920*. New York: Walker, 1970.

Gentile Ford, Nancy. "The Old Country Service School: Gender, Class, and Identity and the YWCA's Training of Immigrant Women in International Social Welfare Philosophy, 1919." *Peace & Change* 23, no. 4 (1998): 440–455.

Gilman, Nils. *Mandarins of the Future: Modernization Theory in Cold War America*. Baltimore, MD: Johns Hopkins University Press, 2004.

Greene, J. P. *The Intellectual Construction of America: Exceptionalism and Identity from 1492 to 1800*. Chapel Hill, NC: University of North Carolina Press, 1993.

Hamer, John. "English and American Giving: Past and Future Imaginings." *History and Anthropology* 18, no. 4 (2007): 443–457.

Hardt, Michael and Antonio Negri. *Empire*. Cambridge, MA: Harvard University Press, 2000.

Hopkins, Charles H. *History of the Y.M.C.A. in North America*. New York: YMCA, 1951.

Horneková, Jana. "Plečnik's Construction of T.G. Masaryk's Apartment," in *Josip Plečnik—An Architect of Prague Castle*, ed. Zeděnek Lukeš and Damjan Prelovšek. Prague: Prague Castle Administration, 1996.

Killen, Linda. "The Rockefeller-Foundation in the 1st Yugoslavia." *East European Quarterly* 24, no. 3 (1990): 349–372.

Kliebard, Herbert M. *Schooled to Work: Vocationalism and the American Curriculum, 1876–1946*. New York: Teachers College Press, 1999.

Krippner, Monica. "The Work of British Medical Women in Serbia during and after the First World War," in *Black Lambs and Grey Falcons: Women Travellers in the Balkans*, ed. John B. Allcock and Antonia Young, 71–89. New York: Berghahn, 2000.

Kroes, Rob. "American Empire and Cultural Imperialism: A View from the Receiving End." *Diplomatic History* 23, no. 3 (1999): 463–477.

Lagemann, Ellen Condliffe. *The Politics of Knowledge: The Carnegie Corporation, Philanthropy, and Public Policy*. 1st ed. Middletown, CT: Wesleyan University Press, 1989.

Lovejoy, Esther Pohl. *Certain Samaritans*. New, rev and reset. ed. New York: The Macmillan Company, 1933.

Low, Brian. "The Hand that Rocked the Cradle: A Critical Analysis of Rockefeller Philanthropic Funding, 1920–1960." *Historical Studies in Education/Revue d'histoire de l'éducation* 16, no. 1 (2004): 33–62.

Masaryk, Alice Garrigue and Ruth Crawford Mitchell. *Alice Garrigue Masaryk 1879–1966: Her Life as Recorded in Her Own Words and By Her Friends.* Pittsburgh, PA: University of Pittsburgh, 1980.

Meyer, John W. and Ronald L. Jepperson. "The 'Actors' of Modern Society: The Cultural Construction of Social Agency." *Sociological Theory* 18, no. 1 (2000): 100–120.

Nolan, Mary. *Visions of Modernity: American Business and the Modernization of Germany.* Oxford: Oxford University Press, 1994.

Reese, William J. "After Bread, Education: Nutrition and Urban School Children, 1890–1920." *Teachers College Record* 81 (1980): 496–525.

Rodgers, Daniel T. *Atlantic Crossings: Social Politics in a Progressive Age.* Cambridge, MA: Harvard University Press, 1998.

Rosenberg, Emily S. and Eric Foner. *Spreading the American dream: American Economic and Cultural Expansion, 1890–1945.* 1st ed., American Century Series. New York: Hill and Wang, 1982.

Sobe, Noah W. "Balkanizing John Dewey," in *Modernities, Inventing The Modern Self and Education: The Traveling of Pragmatism and John Dewey,* ed. Thomas S. Popkewitz, 135–152. New York: Palgrave, 2005.

———. "Cultivating a 'Slavic modern': Yugoslav Beekeeping, Schooling and Travel in the 1920s and 1930s." *Paedagogica Historica* 41, no. 1–2 (2005): 145–160.

———. "Slavic Emotion and Vernacular Cosmopolitanism: Yugoslav Travels to Czechoslovakia in the 1920s and 1930s," in *Turizm: The Russian and East European Tourist under Capitalism and Socialism,* ed. Anne E. Gorsuch and Diane P. Koenker, 82–96. Ithaca, NY: Cornell University Press, 2006.

Szymczak, Robert. "An Act of Devotion: The Polish Grey Samaritans and the American Relief Effort in Poland, 1919–1921." *Polish American Studies* 43, no. 1 (1986): 13–36.

Wagner, Peter. "The Resistance the Modernity Constantly Provokes: Europe, America and Social Theory." *Thesis Eleven* 58 (1999): 35–58.

Wilson, Francesca M. *In the Margins of Chaos; Recollections of Relief Work in and Between Three Wars.* New York: The Macmillan Company, 1945.

Zimmerman, Jonathan. *Innocents Abroad: American Teachers in the American Century.* Cambridge, MA: Harvard University Press, 2006.

Zylberman, Patrick. "Fewer Parallels than Antitheses: René Sand and Andrija Stampar on Social Medicine, 1919–1955." *Social History of Medicine* 17, no. 1 (2004): 77–92.

PART III

Promises of Modernity and Abundance

CHAPTER FIVE

"The Appeal to the German Mind": Educational Reconstruction in the American Zone of Occupation, 1944–49

CHARLES DORN AND BRIAN PUACA

On the afternoon of August 24, 1946, a plane carrying the ten-member U.S. education mission to Germany landed in the war-torn city of Berlin. Chaired by the president of the American Council on Education George Zook and comprised of such nationally renown educators as theologian Reinhold Niebuhr, U.S. Office of Education official Bess Goodykoontz, and president of the George Peabody College for Teachers Henry H. Hill, the mission had been invited by the U.S. State Department to "observe and evaluate" America's effort in reconstructing educational institutions in Germany following World War II. Spending almost four weeks visiting elementary, secondary, and vocational schools, as well as universities in the American zone of occupation, mission members interviewed U.S. military officials, German teachers, pupils, and professors, and toured the burned-out remains of school buildings. The result of their visit was a 50-page report on the U.S. program to rebuild Germany's educational system in the immediate postwar era.[1]

Using the education mission's report as a framework for investigating educational developments in Germany between 1944 and 1949, this chapter examines U.S. officials' attitudes toward education as a component of the postwar reconstruction program. It also inquires into the challenges and dilemmas confronting the U.S. military government as it attempted to reconstruct Germany's education system in a fashion that promoted democratic reform throughout the American zone of occupation. Our conclusion, which interprets the successes and failures of U.S. educational efforts in Germany in light of both reformers' expectations and postwar change, complicates the current historiography by suggesting that curricular and pedagogical innovations in Germany's primary and secondary schools were instrumental to the foundation and growth of democracy in the Federal Republic in the decades following World War II.

Drawing on previously published analyses as well as original archival research, this chapter reveals that U.S. officials considered education a fundamental agent in fostering democracy among a defeated, yet unbowed, German people. From policy statements dictating the need to "control" Germany's educational system, to the governor of the American zone of occupation describing U.S. reconstruction efforts as "the appeal to the German mind," to the education mission's assertion that schools would be "a primary agency for the democratization of Germany," a general consensus existed around the importance of "reeducating" the German people away from fascism and toward democracy.[2] There was much disagreement, however, between government officials, policy makers, and military personnel over the best methods to employ in rebuilding Germany's educational system. Never successfully resolved, these conflicts prevented the United States from both establishing a clearly defined agenda for educational reconstruction and investing the resources necessary to accomplish its ambitious goals during the years of occupation.

As members of the 1946 education mission observed, the United States confronted multiple challenges in Germany at the end of World War II. Having suffered physical devastation resulting from Allied armies invading from both the eastern and western fronts, school buildings across Germany lay in ruins. Textbooks and curricular materials that had not been destroyed during the war were imbued with fascist propaganda and could not be used by pupils. The political loyalties of German educators, many who had been compelled to join the National Socialist Teachers Association (NSLB), had to be determined and those judged pro-fascist "purged" from the system. And the organizational structure of Germany's prewar and wartime schools, which American officials identified as both elitist and authoritarian, required reconstituting. Undeniably a momentous undertaking, the entire process of educational reconstruction was further complicated by a central dilemma—how to resist imposing educational reforms that promised to advance democratic principles on a populace that was not wholly submissive. Writing to U.S. Secretary of State James Byrnes in 1946, Assistant Secretary of State William Benton clearly articulated this dilemma when he noted, "Democracy, by its nature, cannot be imposed. The methods employed by Goebbels, even if we were willing to use them would defeat our purpose. Nevertheless, so long as the United States has the ultimate authority it has the ultimate responsibility to see that the German people work out their own educational salvation."[3] Perhaps more than any logistical hurdle, this dilemma challenged the military personnel responsible for German educational reconstruction in that it compelled the U.S. military government to devise strategies for promoting democratic reforms without resorting to coercion.

Most scholars who have examined American reform efforts in the schools of occupied Germany have judged them to be unsuccessful. This interpretation has come to dominate much of the scholarship. Terms such as *failure* and *restoration* are commonly used to describe the postwar educational system that would emerge in the western zones of occupation.[4] U.S. military historian Harold Zink, political scientist John Gimbel, German historian Karl-Heinz Füssl, and education scholar Gregory Wegner have all questioned the fundamental assumption that the postwar schools were capable of reshaping German pupils in a democratic mold in the chaos of the postwar period.[5] Others have viewed the outbreak of the Cold War in 1947 as signaling the end of any meaningful education reform.[6] In the new Cold War world, education received less attention from American officials interested in rebuilding European economies and containing communism. Nevertheless, a handful of scholars have interpreted the occupation years as a formative period, one that, although not producing clear successes, contributed to positive change over the following decades.[7] This finding underscores the long-term nature of the education reform process and returns agency to the German educators and administrators who were most involved in bringing change to the schools.

The U.S. program to reform German education stands as a case study in the complications of democracy in practice. Perhaps the most important lesson taken from the occupation experience is that democracy cannot be given by one group or taken by another. Democracy is a process that has to be learned. German educators, as evidenced by their actions and writings after the war, formulated their own ideas that adopted, adapted, and sometimes rejected American proposals. German educators formed these ideas about democracy in response to their experiences as citizens of an occupied nation-state, as well as by drawing on German legacies of democracy and reform that dated from Germany's first republic founded following World War I. By developing their own conceptions of democratic education and implementing these ideas during and after the occupation, German educators proved that they understood the basic tenets of American democracy, perhaps even better than many of the American officials charged with "reeducating" German thinking. The greatest legacy of postwar American educational reform, then, may indeed have been the lasting awareness among German educators of the multifaceted nature of democracy.

"That Most Intangible Yet Fundamental of Battlegrounds"

Following the D-Day invasion of Normandy on June 6, 1944, the Allied offensive into Nazi Germany occurred with striking speed. Rapid military advances led the governments of Great Britain, the Soviet Union, and the United States to agree to divide postwar Germany into four zones of occupation (designating a small zone for France). Although establishing a Control Council to make decisions for the German nation as a whole, Allied governments and military commanders claimed significant control over their respective zones. The Americans created an Office of Military Government, United States, (OMGUS) to administer its zone, which was comprised of three German *Länder* or states (Greater Hesse, Württemberg-Baden, and Bavaria), as well as the port city of Bremen. Several large metropolitan areas fell under American control, including Munich, Nuremberg, Frankfurt, Heidelberg, and Stuttgart, and the southwestern sector of Germany's devastated capital, Berlin.

Even prior to Germany's surrender on May 7, 1945, Allied leaders acknowledged that winning the peace in Europe following the end of the war would prove difficult. The Allied failure to implement a lasting security arrangement after World War I led many officials to believe that Nazi military defeat was simply the first step in establishing postwar international stability. The second required "reeducating" the German people. As one scholar observed following the war, "Thoughtful people realized that military victory marked only a phase in a far more basic conflict, one involving economic, psychological, and diplomatic pressures, and one which would ultimately be determined on that most intangible yet fundamental of battlegrounds—the mind of the defeated peoples."[8] Although there was considerable skepticism among the American people about this project, Allied leaders widely shared the hope that Germans could be reeducated away from fascism and toward democracy. General Lucius Clay, for instance, who served as deputy governor of the American zone of occupation from 1945–47 and governor from 1947–49, described the military government's efforts during that period as "the appeal to the German mind."[9] As strong as this hope was, however, it relied on the unproven assumption that reeducation could "psychologically disarm" the German people.[10]

At the Potsdam meeting of Soviet, American, and British leaders in the summer of 1945, Josef Stalin, Harry Truman, and Clement Attlee agreed in principle that "German education shall be so controlled as completely to eliminate Nazi and militaristic doctrines and to make possible the successful development of democratic ideas."[11] Reflecting the Allies' adoption of a primarily punitive approach toward Germany following the war, this multilateral statement also indicated Allied leaders' intention to raise a democratic nation up from the defeat of fascist Germany.[12] The Potsdam statement mirrored U.S. Joint Chiefs of Staff (JCS) directive 1067—the central American occupation directive—which required that "all educational institutions...be closed" and that "a coordinated system of control over German education and an affirmative program of reorientation" be established by the U.S. military government so as to "completely eliminate Nazi and militaristic doctrines and to encourage the development of democratic ideas."[13]

The punitive agenda of JCS 1067, however, conflicted with the strategies of soldiers responsible for implementing German educational reconstruction in the American zone of occupation, making it difficult to establish a clearly defined agenda for rebuilding Germany's educational system. Commissioned into the Education and Religious Affairs (E&RA) subsection of the Allied Expeditionary Force because of their expertise in the fields of education, religion, and culture, E&RA staff issued assessments that frequently contradicted the spirit of JCS 1067. Following the war, for instance, Marshall Knappen, who served as chief of the OMGUS Religious Affairs Section and deputy chief of the Education Section, recalled, "The chief policy issue to be settled was the degree of directness to be used in reorganizing the German educational system."[14] Knappen acknowledged that many "students of the problem" felt that the Americans should "exercise their legal right under military law" to impose reforms on the German people. He and his fellow E&RA staff members, however, strongly believed that "in the long run only the Germans could reeducate themselves; that any plan obviously imposed by foreigners relying almost exclusively on their own judgment would never take root in Germany and would therefore be scrapped the minute our troops were withdrawn."[15]

At the same time that E&RA staff were struggling in Germany to meet JCS 1067 dictates, the National Education Association was urging the U.S. State and War Departments to send a "small committee" of professional educators from the United States to make a first-hand study of both "the remaining evidence of the disastrous effect of Nazi education upon German youth and adults" and "the present situation and outlook for German education."[16] Given the challenges confronting E&RA staff, it is hardly surprising that OMGUS initially resisted the idea of an education mission. John Taylor, the head of E&RA in 1946, thought that such a body could have only one of two objectives: establishing a new educational program or evaluating an existing one. Taylor found neither possibility appealing.[17]

General Douglas MacArthur, however, Supreme Commander for the Allied Powers in Japan, had welcomed the idea of civilians offering assistance in the long-range planning of Japanese reeducation, inviting a group of American educators to Japan in March 1946. Confronted with political pressure following the Japanese mission's well-publicized return to the United States (and not wanting to be unfavorably compared with MacArthur), General Lucius Clay granted access to a group of civilian educators for the purpose of assessing the achievements of the U.S. military government's education program in Germany.[18] Consequently, U.S. Assistant Secretary of State William Benton announced that a mission would travel to Germany "for the purpose of observing and evaluating" the U.S. military government's educational program in the American zone of occupation. "Since the reeducation of the German people from Nazism and militarism toward the acceptance of peace-loving, democratic ideals and ways of life is fundamental for the winning of the peace," Benton observed, "the group of educators will be called upon to bear a responsibility of the highest importance."[19]

Although the education mission to Japan consisted of 27 members, the U.S. military government requested that just 8 educators travel to Germany in August 1946 to conduct the evaluation. As they had with the Japanese mission, State Department officials accepted primary responsibility for recruiting and selecting mission members, which they indicated would include men, women, and representatives of Catholic and Protestant denominations as well as diverse geographic regions within the United States.[20] The U.S. military government, however, expressed dissatisfaction with the State Department's initial list of candidates. Although accepting a proposed two-member increase, it criticized the group for including "no person with teacher training background," "no person with 'general' education (general college) background," and "no person active in public school administration."[21] In response, the State Department modified the list, eventually selecting, inviting, and receiving acceptances from:

— Chairman George F. Zook, President, American Council on Education
— Bess Goodykoontz, Director, Division of Elementary Education, U.S. Office of Education
— Henry H. Hill, President, George Peabody College for Teachers, Nashville, Tennessee
— Paul M. Limbert, President, YMCA College, Springfield, Illinois
— Earl J. McGrath, Dean, University of Iowa, Iowa City, Iowa
— Reinhold Niebuhr, Professor, Union Theological Seminary, New York, New York
— Reverend Felix Newton Pitt, Secretary, Catholic School Board, Louisville, Kentucky
— Lawrence Rogin, Director of Education, Textile Workers Union of America, CIO, New York
— T.V. Smith, Professor, University of Chicago, Chicago, Illinois
— Helen C. White, Professor, University of Wisconsin, Madison, Wisconsin[22]

The group, which was comprised primarily of professional educators who had dedicated their careers to elementary, secondary, or higher education teaching and/or administration, shared a progressive vision of schooling's social function that was both rooted in early twentieth-century liberal educational theory and characteristic of the American educational elite during the war era.[23] Describing their intentions for Germany's postwar educational system, for instance, mission members employed language characteristic of the progressive education movement to define schools' role in fostering democratic attitudes:

> The school emerges as the common center of mutuality, where ideally all children meet all children as fellow-children before any have been narrowed by class or creedal bias. But even to approach this ideal we must have not merely the essentially negative safeguards of creed, race, and class toleration but have also exemplified in the school the positive *method* of living which democratic

citizenship enshrines and climaxes [emphasis original]. The goal of democracy is the democratic man.[24]

Reflecting Deweyian principles of education, the mission's description of the school as a laboratory of democracy represented progressive education's "conventional wisdom," to borrow historian Lawrence Cremin's classic phrase, and offered mission members a common discourse for use in conducting its evaluation.[25]

The education mission arrived in Berlin on August 24, 1946, and after spending several days conferring with U.S. military government officials traveled throughout the American zone of occupation. The U.S. military government compiled much of the data that mission members examined prior to their arrival. Nevertheless, the mission visited German schools, colleges, universities, and informal educational groups and held meetings with American military officials and E&RA staff as well as with German teachers, pupils, and professors. Although their observations were circumscribed by time limitations, members reported substantive discussions with German educators regarding school and community needs.[26] Following their return to Berlin, mission members spent almost one-fourth of their total time in Germany conferencing, analyzing, and interpreting their observations. Chairman Zook submitted the mission's final report prior to departing for the United States on September 26. On October 15, 1946, the State Department released 30,000 copies of the evaluation for distribution in the United States and another 20,000 copies of its German translation for distribution in Germany.[27]

"A Task Contradictory to Democratic Genius"

Although the mission's report began with an extensive overview of the development of German schooling—what members labeled "Factors Conditioning German Education"—neither of the two factors identified as bearing influence on the nation's educational system involved Nazi ideology.[28] Indeed, mission members observed that by the end of the war, "The more perverse and obvious Nazi theories and practices..." had already become "abhorrent to the German people themselves." Instead, the mission held "the special character of German culture, with its peculiar defects and virtues" responsible for inhibiting German democratic development, especially in educational institutions. Conceiving of German history and culture as encompassing a central paradox, the mission reported, "No country—unless it be ancient Greece or Rome—has contributed more generously to the common treasures of our civilization." Yet mission members also noted, "some of the deeper sources of recent perversities are to be found in this same culture...some of the perversions of Nazism were exaggerations of strains of thought and feeling, deeply rooted in German history."[29]

The second factor mission members credited with conditioning Germany's educational system resulted directly from the war—political, economic, social, and physical devastation. "Nowhere in the world," they declared, "has it been possible to erect the structure of successful democratic self-government upon starvation and disorder."[30] The mission claimed that for German "reeducation" to succeed, "it is necessary that an economy exist, or be in the making, in which the democratic spirit can develop and democratic institutions be established."[31] This challenge was complicated not only by the general disorder associated with Germany's military defeat but by its division into zones of occupation, which in the summer of 1946 were already proving to be a point of contention between the Western Allies and the Soviet Union. Moreover, two million German refugees from Nazi annexed territories in Czechoslovakia and Hungary returned to the American zone of occupation following the war, making an already problematic situation especially difficult.

Locating adequate facilities for classroom instruction proved to be one of the most pressing challenges for American officials in the first months of the occupation. Conditions were particularly bleak in urban areas, where the bombing had been most severe. Munich serves as a telling example, with 10 percent of the city's schools totally destroyed and 78 percent in need of renovation before classes could resume.[32] The situation was much the same in Berlin, which had also bore the brunt of Allied bombing raids. Of the 608 schools existent in Berlin before the war, 124 were totally destroyed and another 111 were in need of serious repair before they could be reopened.[33] In the American sector of the city, the schools in Steglitz, Kreuzberg, Schöneberg, and Neukölln were hardest hit, although the situation in other districts was not much better.[34] The circumstances were admittedly less dire outside the major urban areas in the American zone, however, there were few places where the schools escaped unscathed from the ravages of war.[35]

Adding to the crippling effects of wartime bombings were the requisitions of useable schools for noneducational purposes. Both American military officials and German authorities occupied school buildings that were still standing after May 1945. Although appropriated for a variety of purposes, the most common uses were: medical facilities, office space for occupation officials and German governmental agencies, and housing for displaced persons. This further complicated the postwar situation, since the schools that were ordinarily housed in these facilities had to locate other classroom space. In Berlin, the occupation of 81 schools for noneducational purposes further restricted the available classroom space in the city.[36] These confiscations, in addition to the schools that were unusable due to bombing, meant that only 3,044 classrooms—approximately 23 percent of the prewar number—were available for instruction in the fall of 1945.[37] The situation was arguably less dire elsewhere in the American zone, particularly outside the cities, yet locating classroom space would continue to be a challenge for OMGUS officials throughout the occupation.

Another obstacle to the reopening of the schools was the combination of material shortages and disease. Teachers complained of a lack of basic classroom materials, including desks, pencils, and slates. Particularly problematic was the scarcity of paper, which restricted instruction in the classroom as well as the amount of homework assigned. Shortages of other basic necessities—shoes, clothes, and coats—also prevented many pupils from traveling to school in inclement weather.[38] Additionally, disease and malnourishment forced thousands of pupils to miss school. For example, an outbreak of tuberculosis in Neukölln, one of the American-occupied districts in Berlin, kept hundreds of pupils away from classes just as their teachers were welcoming them back for the first time in several months.[39] As autumn turned to winter, a shortage of coal necessary for heating the schools provided American officials with yet another challenge. Nevertheless, by October 1, 1945, E&RA staff had secured sufficient space and school supplies to permit reopening elementary schools in the American zone of occupation. By year's end, approximately 1.8 million pupils—80 percent of Germans between the ages of 6 and 14—were attending class.[40]

In the report's second section, mission members described a number of tensions they encountered while conducting their evaluation. The first involved the dilemma of how to promote cooperative, democratic principles in German educational reconstruction through the use of coercive, undemocratic methods. "Military victory," the members reported, "has committed the Allies to a task contradictory to democratic genius. This genius is to allow and even to encourage variety in thought and feeling. Far from this luxury, however, military success has obligated the democracies not only to an untoward recommendation of their own virtues but to a downright denial of the right of Axis ideologies further to propagate or even longer to live."[41]

Dissatisfied with this dilemma but failing to propose alternative approaches, the mission chose not to criticize OMGUS for its punitive policies but instead relied on the language of the Potsdam accords as a standard against which to judge U.S. efforts in reconstructing German education. Among the punitive policies that had sparked the greatest ire among the Germans were the widespread removals of experienced educators classified as Nazis and the improvised censoring of classroom texts by OMGUS officials. Understanding the Potsdam agreement as consisting of a "double directive" comprised of a "negative and positive aspect" of America's "commitment to victory," mission members interpreted the negative aspect—"German education shall be so controlled as to completely eliminate Nazi and militaristic doctrines"—as justifying the military government's "initial severity" in denazifying teachers and textbooks.[42]

At the heart of the American mission to reform the German schools was the purging of Nazi educators from the school. Denazification, as the process came to be called, prompted the removal of all teachers deemed to be "Nazis" from their positions. Defining who was a Nazi in a country in which more than 90 percent of educators had been compelled to join the Nazi professional organization for teachers, however, proved to be an unprecedented challenge.[43] Due to both a fluid definition of what constituted a "Nazi teacher" and the decentralized nature of the denazification process, removal rates varied dramatically throughout the American zone. American officials designed a detailed questionnaire, or *Fragebogen*, that was supposed to standardize the denazification process and allow for a complete analysis of the political activities of individuals. Even with this tool, OMGUS officials applied policy in an uneven and inconsistent fashion throughout the American zone. For large regions, the best estimates for summer 1945 suggest that denazification proceedings removed 35–40 percent of the teaching corps.[44] The purging of German teachers by American officials would actually intensify in 1946. As renewed fears of a renewed Nazi presence in the schools prompted heightened vigilance among OMGUS officials, removal rates in many areas exceeded 50 percent. The result of these policies was precisely what many Americans had feared: pupil-to-teacher ratios that exceeded 60:1 in many areas.

The education mission took issue with the "rough and ready procedures" used to screen the German teaching force. According to mission members, the military government removed from their positions many teachers who had only a passing interest in the Nazi Party, creating an immense teacher shortage. They also asserted that no productive system of democratic education could be developed with pupil-to-teacher ratios at such elevated levels and with pupils attending class for only two hours before another shift of young Germans arrived to be taught by the same teacher (*Schichtunterricht*). To prevent such missteps in the future, the mission urged devolving control over educational reconstruction to German civil authorities and recommended that "The respective [German] Länder ministries should be allowed to screen teachers whose dismissal was never mandatory and to reemploy at once on probationary status those found to be at once least politically unfit and most efficient pedagogically."[45] While this advice certainly made sense from an educational standpoint, it also exposed OMGUS officials to pressure from an American public sensing leniency in dealing with former Nazis.

Providing a reasonable supply of textbooks for use in German schools also posed a significant problem for E&RA staff. There was no question that textbooks published after the Nazi seizure of power in 1933 were unacceptable for continued use in the classroom. OMGUS officials identified objectionable material in virtually all subject areas—history, mathematics, languages, and the sciences. These books were infused with anti-Semitism, extreme nationalism, xenophobia, and outright Nazi propaganda. Unable to delete Nazi bigotry from the texts and not having the luxury of time to contract for new texts, the U.S. military sought an adequate supply of pre-Nazi textbooks published during the Weimar Republic.

As early as 1944, American officials in Aachen had prepared temporary "new" texts for use in the schools under American control.[46] With an initial printing of a paltry 40,000 copies, these books, which were reproductions of editions in the collections of Columbia University Teachers College, were designed to serve as useable texts until new ones could be printed on a large-scale basis. Refusing to allow German émigrés to the United States to author new textbooks, OMGUS officials bolstered their case that they were not trying to "Americanize" the German schools. Yet these Weimar era textbooks also displayed nationalistic, militaristic, and xenophobic content that American officials and many German educators deemed unacceptable for use in the postwar classroom. Thus American officials spent much of 1945 reviewing and vetting these texts in order to produce an adequate supply of books for the first phase of the occupation. By the end of 1945, American education officials had vetted over five million texts and approved them for use in the schools of the zone. There were 20 volumes in total: 8 readers, 5 arithmetic books, 3 history texts, and 4 editions devoted to the natural sciences.[47] These "emergency textbooks" did not meet the needs of all pupils, and they were by no means a permanent solution, but they would serve as the basis of instruction in many schools through 1948.[48]

"The Successful Development of Democratic Ideas"

Turning its attention to the positive aspect of the Potsdam accord—"the successful development of democratic ideas"—the education mission issued a harsh criticism of the organization and philosophical underpinnings of Germany's educational system. Historically, Germany's schools and curricula were rigidly tracked. By World War II, most children attended some preschool, or *Kindergarten*, and beginning at age five or six lower elementary school, or *Grundschule*. In most areas, this common elementary school experience lasted only four years. At the age of nine or ten children underwent examination for admission to secondary school. Ten percent of pupils gained entrance to institutions intended to prepare them for higher education, including the *Gymnasium*, which offered a classical curriculum, and the *Oberrealschule*, which offered a curriculum emphasizing sciences and modern languages. The *Aufbauschule*, a school designed to provide rural pupils and "late bloomers" a chance to "catch-up," allowed access to higher education or vocational training tracks.[49] The other 90 percent of pupils proceeded to the upper grades of elementary school, known as the *Volksschule*, or to the *Realschule* for eventual business, technical or other vocational training. When these pupils turned 14 or 15, they began four or more years of full-time vocational education (divided between classroom and practical training) in the *Berufsschule*, or full- or part-time vocational training through the *Fachschule*. The outcome of this organizational structure was a system of education that determined, to a significant degree, pupils' educational and career opportunities at a strikingly early age.[50]

Perceiving the school as "a primary agency for the democratization of Germany," mission members claimed in their report that this system "cultivated attitudes of superiority in one small group and inferiority in the majority of the members of German society, making possible the submission and lack of self-determination upon which authoritarian leadership has thrived." They continued, "Nowhere is there the possibility of a common school life, nor in fact any place for that broad base of general education which in many other countries provides a common cultural and social experience.... It is clear that the educational system of a country may reenforce the foundations of a 'class society,' or it may build a common culture for all citizens. For a democratic society, the second is the only possible choice."[51]

In place of the traditional system, the mission proposed an American-style educational system to foster democratic education in Germany. Accordingly, the mission recommended uniting the nation's elementary, secondary, and vocational schools into one comprehensive system. "The terms 'elementary' and 'secondary' in education," mission members suggested "should not primarily be conceived as meaning two different *types* or *qualities* of instruction... but two consecutive levels of it, the elementary one comprising grades 1 to 6, the secondary one those from 7 to 12. In this sense the *vocational* schools should be considered a part of the secondary school system" [emphasis original].[52]

Mission members expected this typically American educational arrangement to promote democratic values over what they perceived as authoritarian and elitist ones emphasized in German schools. They also noted, however, one significant difference between their proposed system and America's public schools—German law provided for no separation of church and state. As in the United States, the German school system had historically been separated into private parochial schools, or *Bekenntnisschulen*, and public community schools, or *Simultanschulen*. In Germany, however, both school systems received state support. Moreover, pupils attended religious classes in both systems, with teachers from either Catholic or Protestant faiths instructing public school pupils with similar denominational affiliations.[53]

The mission credited "counsels of prudence no less than considerations of humility" with preventing the U.S. military government from imposing separation of church and state on Germany following the war.[54] Yet mission members also noted that it should not "in the name of democracy allow such arrangements in education as will deprive any religious claimants of equal opportunity or as will through continuous bickering for pious advantage depreciate the high claims of a free spiritual life upon the very generation whose magnanimity will condition the future of the democratic way of life in Germany."[55] In other words, although willing to defer to tradition in this area of German educational life, mission members privileged developing democratic qualities over religious instruction in Germany's schools. "Toleration," they concluded, "must be guaranteed not only for different believers but for disbelievers as well, and the school must be maintained as a place where the young have at least a chance to grow less sectarian than the old."[56] As sternly as it asserted this position, however, the mission recommended neither alternative structural arrangements for religious education nor guidance on how to provide a democratic education in a sectarian learning environment.

Unable to resolve the tension between educating for democracy and religious belief in German education, the mission targeted the school curriculum as the area with greatest potential for fostering democratic values. In contrast to the significantly different academic and vocational curricular tracks historically experienced by German pupils, the mission emphasized the need for the "whole school program" to "make a significant contribution to democratic experience."[57] Using strikingly progressive language, the mission described the German secondary school curriculum as "crowded with subjects, heavy with academic tradition, and in most respects remote from life and ill-adapted to the present and future needs of the pupils."[58] It recommended replacing this program with "an elastic organization of the curriculum in core subjects and elective courses," permitting "the differentiation necessary with regard to the future vocational or professional intentions of the students."[59] The "most important" component of this reform, according to the mission, involved instruction in democratic citizenship, especially in the "content and form" of the social sciences. Mission members believed that history, geography, and civics would play crucial roles in nurturing democratic attitudes and values among German pupils. They insisted, moreover, that democratic principles infuse German teaching methods, including "cooperative class projects, classroom committees, discussion groups, school councils, student clubs," and "community service projects."[60] This final recommendation promoting the cultivation of miniature democratic communities in the schools had a great impact on German educators. Although the schools did initiate the ambitious reforms promoted by OMGUS officials during the occupation, their impact would only begin to become visible in the years after the Americans' departure.[61]

Following their evaluation of German elementary and secondary schools, mission members turned their attention to the universities and "higher schools," teacher education, youth groups, adult education, films, radio, and libraries, highlighting the educational qualities of these areas as well as the contributions each could make to developing democracy in Germany. Among its recommendations, the mission suggested a longer period of teacher training and establishing "a separate pedagogical faculty" at German universities "for the teaching of the professional subject matter required by the future teachers."[62] Likewise, the report emphasized the need for elementary and secondary educators to be educated together at the university level in order to eradicate professional tensions and to ensure all teachers received the very best preparation for their careers.[63] As with elementary and secondary education, the mission claimed that denazifying German college and university staff was a necessary but not adequate condition for developing democratic approaches to higher education. "Instruction must be provided which will inform students about domestic and international affairs and teach them the habits and techniques of democratic living," mission members reported. "It is recommended that all universities and higher schools include within each curriculum the essential elements of general education for responsible world citizenship and for an understanding of the contemporary world."[64]

Youth groups had played a highly visible role in Nazi propaganda during the 1930s and, according to the mission, had "special significance for the reeducation of the German people."[65] mission members affirmed their belief that "the attitudes and ideas of young people may be modified more readily than those of their elders in the direction of a democratic way of living" and suggested that the military government continue to encourage German youth to form voluntary associations "to provide for a constructive use of leisure time and to give training in democratic organization and procedures."[66] Similarly, the mission proposed establishing education programs to provide adults with "training in self-government to give an understanding of its spirit as well as its techniques of operation."[67] It also recommended providing adults with accurate information on the state of "world affairs" through adult education programs, by using film and radio, and by replacing library books burned during the Nazi era.[68]

The mission's report ended with a host of long-term recommendations. They sought to significantly increase the number of E&RA personnel, to continue American aid for reconstructing Germany's educational system, and to develop an extensive exchange program between the United States and Germany for "students, teachers, and other cultural leaders."[69] Mission members also reaffirmed their essential faith in the power of education to catalyze change in Germany's political, economic, and social systems. "For this process of attaining democracy in Germany in this generation," they wrote, "education is the one best instrument to employ.... Hence, so long as the United States continues as an occupying power in Germany, it should encourage and use the instruments of education to attain its major purpose, namely the attainment both of a democratic spirit and form of government."[70]

Conclusion

Writing for *The Washington Post* one year following the education mission's return to the United States, Fred M. Hechinger declared that the German people had "glibly" rejected "American school reforms." "Today it is clear," Hechinger wrote, "that the run of the mill German education official, from the Minister down, does not intend to carry out any large-scale reforms whatsoever. At best, in the more progressive sections of the country, minor concessions have been made."[71] Many scholars have concurred with Hechinger's pessimistic assessment of American efforts to reconstruct Germany's educational system following World War II. Regarding the mission's evaluation specifically, historian Arthur Hearnden has written, "in the end their [mission members'] plans for restructuring the system made very little impact."[72] Historian Gary Tsuchimochi has similarly noted that the mission had "little effect on German educational reform."[73]

The U.S. education mission report addressed an American audience when it was released in October 1946 and, as a consequence, was read by few German educators or officials. To argue, however, that American efforts to reform the German schools were wholly unsuccessful is unjustified. The report highlighted areas on which OMGUS officials would concentrate their efforts for the duration of the occupation. Notable progress was achieved in the areas of facilities, supplies, and teacher training in the 2 years following the publication of the report. Likewise, the year 1949 might have marked the close of the occupation and the birth of West Germany, but it did not signify the end of reform in the schools. Endorsing such a long-term view of the occupation and its effects, James Tent has argued that the "evolution of education values in the Federal Republic since 1949 has undoubtedly benefited from the education-reform efforts of the Americans, British, and French."[74] The West German schools continued to experience important reforms throughout the next decade, which were at least in part sparked by the occupation.

As Thomas Koinzer demonstrates in this volume, long-term democratic educational reform became visible several decades after the Americans had returned home. A much more immediate tension with which Americans grappled was the paradox of how to implement democratic educational reforms without resorting to undemocratic methods such as coercion, imposition, and control. The education mission attempted to resolve this dilemma by acknowledging the "stern and corrective task" comprising America's occupation and pacification of

Germany and issuing its approval of the military government's dena-zification program.[75] However, when addressing German educational reconstruction, mission members recommended "the most complete possible civilianization of educational authority and of military govern-ment and in the transference of functions of authority to German agen-cies of government."[76] In other words, rebuilding Germany's political and economic structures might necessitate long-term military intrusion, but reconstructing Germany's educational institutions required trans-ferring authority away from coercive American military leaders toward cooperative German civilians. "This transfer of authority is a wise mea-sure," the mission urged, "creating democratic life by practice rather than precept."[77] This was indeed a prophetic statement. Many of the reforms promoted by American officials during the occupation would be realized in the following decades precisely because they were under-taken by German educators who adapted them for their own uses.

Members of the education mission trusted that the German people would adopt democratic educational reforms of their own free will. Doing so, according to the mission, would personally engage Germans in the prac-tice of democracy. Germany's failure to embrace democratic educational reforms during the Weimar Republic following World War I, however, suggests the extent to which the mission's optimism was ungrounded. Having uncovered a conservative ideology infusing Germany's educa-tional system, mission members qualified their own recommendations, urging that the U.S. military government's "advisory function" be "sup-plemented by the right of veto, to be exercised whenever our stated objec-tives of developing a democratic education seemed imperiled."[78]

Unable to resolve the tension between cooperation and control, the education mission left the U.S. military government without an alterna-tive approach to German educational reconstruction. When Germans in the American occupation zone resisted U.S. proposals for the creation of comprehensive secondary schools in the years following World War II, the military government, lacking clear guidance from the mission's evaluation, continued to focus on solving basic problems such as a shortage of teachers, buildings, textbooks, and supplies. As a result, the three-tiered organiza-tion of Germany's education system remained essentially unaltered in the postwar era. This, however, did not prevent the implementation of mean-ingful change in classrooms throughout the American zone. The German experience underscores the centrality of curricular and pedagogical reforms to the rebuilding of education in postwar states. Democracy cannot be imposed from without; rather, it has to be experienced and accepted from within. The American occupation provided precisely this opportunity for

Germany's schools. While it would be wrong to refer to OMGUS officials as the architects of German democracy, they certainly could be considered the draftsmen who helped give life to the project.

Notes

Excerpts above from Charles Dorn, "Evaluating Democracy," *American Journal of Evolution* 26, no. 2 (2005): 267–277 reprinted by permission of Sage Publications.

1. "Report of the United States Education Mission to Germany," (Washington, DC: Department of State, 1946).

2. Educational reform in the postwar German schools was not solely a Western project. The Soviets sought to create "new schools" in their zone of occupation that would promote antifascist democratic education. At the center of these reforms in the Soviet zone was the *Einheitsschule*, which was a comprehensive secondary school—with German origins—that offered a traditional curriculum. See Benita Blessing, *The Antifascist Classroom* (New York: Palgrave Macmillan, 2006).

3. Quoted in Masako Shibata, *Japan and Germany under the U.S. Occupation: A Comparative Analysis of Post-War Education Reform*, ed. Edward R. Beauchamp, *Studies of Modern Japan* (Lanham: Lexington Books, 2005), 118.

4. There is great consensus as to the "restoration" of pre-Nazi education after 1945. See Saul Robinsohn and J. Caspar Kuhlmann, "Two Decades of non-reform in West German Education," in *Education in Germany. Tradition and Reform in Historical Context*, ed. David Phillips (New York: Routledge: 1995); Arthur Hearnden, *Education in the Two Germanies* (Boulder, CO: Westview, 1974); Jutta-B. Lange-Quassowski, *Neuordnung oder Restauration? Das Demokratiekonzept der amerikanischen Besatzungsmacht und die politische Sozialisation der Westdeutschen* (Leske Verlag: Opladen, 1979).

5. Harold Zink, *The United States in Germany, 1944–1955* (New York: D. Van Nostrand Company, 1957); John Gimbel, *A German Community Under American Occupation: Marburg, 1945–1952* (Stanford, CA: Stanford University Press, 1961); Karl-Heinz Füssl and Gregory Wegner, "Education Under Radical Change: Educational Policy and the Youth Program of the United States in Postwar Germany," *History of Education Quarterly* 36, no. 1 (1996): 6.

6. Jutta B. Lange-Quassowski argues that German education had not in fact begun before 1947, as denazification had prevented its implementation. Karl-Heinz Bungenstab offers a different interpretation, claiming that 1947 marked the end of positive reforms because the United States shifted its priorities to containing communism. Henry Kellermann argues that reform, however limited through 1949, was not implemented based on Cold War ideological motives. See Jutta-B. Lange-Quassowski, "Amerikanische Westintegrationspolitik, Re-education and deutsche Schulpolitik," in *Umerziehung und Wiederaufbau: Die Bildungspolitik der Besatzungsmächte in Deutschland und Österreich*, ed. Manfred Heinemann (Stuttgart: Klett-Cotta, 1981), 53–67; Karl-Ernst Bungenstab, *Umerziehung zur Demokratie? Reeducation-Politik im Bildungswesen der US-Zone 1945–1949*, (Gütersloh: Bertelsmann Universitätsverlag, 1970); Henry Kellermann, "Von Re-education zu Re-orientation: Das amerikanische Re-orientierungsprogramm im Nachkriegsdeutschland," in Heinemann, 86–102.

7. See James Tent, *Mission on the Rhine: Reeducation and Denazification in American-Occupied Germany* (Chicago, IL: University of Chicago Press, 1982); H.J. Hahn, *Education and Society in Germany* (New York: Berg, 1998), 91–112. Karl-Heinz Füssl also alludes to some of the long-term influences of the American occupation in his important study of postwar education in all four of the Allied zones. See *Die Umerziehung der Deutschen. Jugend und Schule unter den Siegermächten des Zweiten Weltkriegs 1945–1955* (Paderborn: Schöningh, 1995).

8. Robert King Hall, "The Battle of the Mind: American Educational Policy in Germany and Japan," *Columbia Journal of International Affairs* 2, no. 1 (1948), 59.

9. Lucius D. Clay, *Decision in Germany* (New York: Doubleday & Co., 1950), especially chapter fifteen.

10. *Occupation of Germany: Policy and Progress*, (Washington, DC: Department of State, Office of Public Affairs, Publication 2783, European Series 23, 1947), 61.

11. *Germany, 1947–1949: The Story in Documents*, (Washington, DC: Department of State, Office of Public Affairs, Publication 3556, European and British Commonwealth Series 9, 1950), 49.

12. Conflicts between the U.S. Treasury, State, and War Departments over the character of America's occupation policy in Germany proved contentious throughout the occupation period, however. In 1944, Treasury Secretary Henry J. Morgenthau, a personal friend of President Roosevelt's, had convinced the president to agree to a harshly punitive agenda for the postwar reconstruction of Germany. The *Morgenthau Plan*, as it came to be called, involved deindustrializing the Ruhr and Saar regions of Germany and dividing the nation into multiple "pastoral" states. State Department officials, on the other hand, adopted economic restoration as a central priority, leading to their favoring a set of occupation policies that emphasized "democracy building" and cooperation with the German people in ridding their nation of fascism. For its part, the War Department desired as little civilian interference as possible in the military occupation of Germany. Although negative publicity in the context of an election year led Roosevelt to withdraw his support for the Morgenthau Plan, he resisted approving an alternative policy for the American occupation. "Speed on these matters is not essential at the present moment," Roosevelt told Secretary of State Cordell Hull in the fall of 1944. "I dislike making detailed plans for a country which we do not yet occupy." Indeed, even in April 1945, with Germany's surrender imminent, Roosevelt claimed, "our attitude [toward German occupation] should be one of study and postponement of final decision." Quoted in Tent, *Mission on the Rhine*, 30. Also see Carolyn Eisenberg, *Drawing the Line: The American Decision to Divide Germany* (Cambridge: Cambridge University Press, 1996), 14–70; John Lewis Gaddis, *The United States and the Origins of the Cold War, 1941–1947* (New York: Columbia University Press, 1972), 96–97; Marc Trachtenberg, *A Constructed Peace: The Making of the European Settlement, 1945–1963* (Princeton, NJ: Princeton University Press, 1999), 15–17.

13. *Germany, 1947–1949*, 26. JCS 1067 served as the only approved U.S. policy statement on the American occupation until July 1947 when it was superseded by the more moderate and less vengeful JCS 1779. Henry J. Kellermann, *Cultural Relations as an Instrument of U.S. Foreign Policy: The Education Exchange Program between the United States and Germany, 1945–1954* (Washington, DC: U.S. Department of State Bureau of Educational and Cultural Affairs, 1978), 19. Also see John L. Snell, *Dilemma over Germany* (New Orleans, LA: Hauser Press, 1959), 176–191; Walter L. Dorn, "The Debate over American Occupation Policy in Germany in 1944–1945," *Political Science Quarterly* 72, no. 4 (1957).

14. Marshall Knappen, *And Call It Peace* (Chicago, IL: University of Chicago Press, 1947), 44.

15. Ibid., 45.

16. Quoted in Gary H. Tsuchimochi, *Education Reform in Postwar Japan: The 1946 Education Mission* (Tokyo: University of Tokyo Press, 1993), 177. Although Tsuchimochi's focuses on developments on postwar Japan, he offers a valuable comparison of the Japanese and German education missions. See Tsuchimochi, *Education Reform in Postwar Japan*, especially chapter seven.

17. Tent, *Mission on the Rhine*, 112–113; Shibata, *Japan and Germany under the U.S. Occupation*, 118.

18. Tent, *Mission on the Rhine*, 112–113. The purpose of the education mission to Japan was "to assist and advise" the U.S. military's education program in that nation. In contrast, the purpose of the mission to Germany was to observe and evaluate. Whether mission members were to assess the assumptions underlying the U.S. military government's educational program in Germany or simply evaluate its success in achieving a set of previously determined objectives

was never precisely clarified, leading to some confusion over the mission's actual role. On U.S. effort to reconstruct Japanese education following World War II, see chapter 6 in this volume.

19. "Report of the United States Education Mission to Germany," xiii.
20. Tsuchimochi, *Education Reform in Postwar Japan*, 183.
21. Quoted in ibid., 183.
22. "Report of the United States Education Mission to Germany," xiii.
23. Progressive educational theory also infused the reconstruction of the Italian educational system following World War II. See, for instance, Allemann-Ghionda, Cristina. "Dewey in Postwar-Italy: The Case of Re-education," *Journal Studies in Philosophy and Education*. 19, no. 1, (2000), 53–67 and White, Steven F. *Progressive Renaissance: America and the Reconstruction of Italian Education, 1943–1962*. New York: Garland, 1991.
24. Ibid., 14.
25. Lawrence Cremin, *The Transformation of the School: Progressivism in American Education, 1876–1957* (New York: Vintage Books, 1961), 32. Given space constraints, this essay will neither examine progressive education's historical development nor attempt to define its various strands. Numerous educational historians have conducted such analyses, however. See David Tyack, *The One Best System: A History of American Urban Education* (Cambridge, MA: Harvard University Press, 1974); Larry Cuban, *How Teachers Taught: Constancy and Change in American Classrooms, 1890–1990*, 2nd ed. (New York: Teacher's College Press, 1993); Arthur Zilversmit, *Changing Schools: Progressive Education Theory and Practice, 1930–1960* (Chicago, IL: University of Chicago Press, 1993); Herbert M. Kliebard, *The Struggle for the American Curriculum, 1893–1958*, 2nd ed. (New York: Routledge, 1995); Diane Ravitch, *Left Back: A Century of Failed School Reform* (New York: Simon and Schuster, 2000).
26. "Report of the United States Education Mission to Germany," 6.
27. Tsuchimochi, *Education Reform in Postwar Japan*, 196.
28. "Report of the United States Education Mission to Germany," 1.
29. Ibid., 1.
30. Ibid., 4.
31. Ibid., 5.
32. OMGUS, *Education and Religion—Monthly Report of Military Governor*, no. 1, July 1945, microfiche, 2.
33. "Die Berliner Schulen vor und nach dem Kriege," Letter from Magistrat von Gross-Berlin (Hauptschulamt), February 13, 1948, C. Rep. 120, No. 3304, Landesarchiv Berlin (LAB).
34. "School Survey—U.S. Sector, Berlin," Box 75; Records Relating to Cultural Exchange and School Reopenings; Records of the Education Branch; Records of the Education and Cultural Relations Division; Records of the U.S. Occupation Headquarters, WWII, Record Group 260, National Archives, College Park (NACP).
35. Brian M. Puaca, *Learning Democracy: Education Reform in West Germany, 1945–1965* (New York: Berghahn Books, 2009), 25.
36. Ibid, 26.
37. "Die Berliner Schulen vor und nach dem Kriege," C. Rep. 120, No. 3304, LAB. According to American statistics, only 20 of the 81 schools in Berlin used for noneducational purposes were located in the American sector.
38. Puaca, *Learning Democracy*, 27.
39. Dietmar Schiller, "Schulalltag in der Nachkriegszeit," in *Schulreform—Kontinuitäten und Brücke. Das Versuchungsfeld Berlin-Neukölln*, vol. 2, Gerd Radde and Werner Korthaase, et al. eds. (Opladen: Leske & Budrich, 1993), 35.
40. Knappen, *And Call It Peace*, 86–87.
41. "Report of the United States Education Mission to Germany," 10.
42. Ibid., 10.
43. According to American statistics, the 98 percent of teachers—totaling 360,000 individuals—belonged to the NSLB by 1942. See "Supreme Headquarters Allied Expeditionary Force

G-5 Division Education and Religious Affairs Handbook," February 17, 1945, Entry 16, File 129966; Box 1497; Intelligence Reports (XL Series), 1941–1946; Research and Analysis Branch Divisions; Records of Office of Strategic Services, RG 226, NACP.

44. Tent, *Mission on the Rhine*, 53.
45. "Report of the United States Education Mission to Germany," 11.
46. Education and Religious Affairs Branch, *Textbooks in Germany-American Zone*, August 2, 1946; Box 702; Education Reports of the Education Branch, 1947–1948; Education and Cultural Relations Division; Records of Office of Military Government, Hesse; Records of the U.S. Occupation Headquarters, WWII, RG 260, NACP.
47. Puaca, *Learning Democracy*, 29–30.
48. "Memorandum on the Selection of Pre-1933 German Texts," undated; Box 78; Records Relating to Cultural Exchange and School Reopenings; Records of the Education Branch, Records of the Education and Cultural Relations Division; Records of the U.S. Occupation Headquarters, WWII, RG 260, NACP.
49. Tent, *Mission on the Rhine*, 116.
50. Ibid., 116; Hearnden, *Education in the Two Germanies*, 30–31.
51. "Report of the United States Education Mission to Germany," 19.
52. Ibid., 22.
53. The mission estimated that 96 percent of Germans living in the American zone of occupation belonged to either Catholic or Evangelical denominations. Ibid., 20.
54. Ibid., 13.
55. Ibid., 13.
56. Ibid., 13.
57. Ibid., 22.
58. Ibid., 22.
59. Ibid., 22.
60. Ibid., 23.
61. See Puaca, *Learning Democracy,* 193–197.
62. "Report of the United States Education Mission to Germany," 26, 31.
63. "Report of the United States Education Mission to Germany," 23.
64. "Report of the United States Education Mission to Germany," 30.
65. Ibid., 33.
66. Ibid., 33–34.
67. Ibid., 36.
68. Ibid., 36–40.
69. Ibid., 44–47.
70. Ibid., 50.
71. Fred M. Hechinger, "Germans Glibly 'Veto' American School Reforms," *The Washington Post*, September 28, 1947, B8.
72. Hearnden, *Education in the Two Germanies,* 37.
73. Tsuchimochi, *Education Reform in Postwar Japan*, 200.
74. Tent, *Mission on the Rhine*, 318.
75. "Report of the United States Education Mission to Germany," 2.
76. Ibid., 2.
77. Ibid., 43.
78. Ibid., 43.

Bibliography

Allemann-Ghionda, Cristina. "Dewey in Postwar-Italy: The Case of Re-Education." *Journal Studies in Philosophy and Education.* 19, no. 1, (2000), 53–67.

Blessing, Benita. *The Antifascist Classroom: Denazification in Soviet-occupied Germany, 1945–1949.* New York: Palgrave Macmillan, 2006.

Bungenstab, Karl-Ernst. *Umerziehung zur Demokratie? Reeducation-Politik im Bildungswesen der US-Zone 1945–1949.* Gütersloh: Bertelsmann Universitätsverlag, 1970.

Clay, Lucius D. *Decision in Germany.* New York: Doubleday & Co., 1950.

Cremin, Lawrence. *The Transformation of the School: Progressivism in American Education, 1876–1957.* New York: Vintage Books, 1961.

Cuban, Larry. *How Teachers Taught: Constancy and Change in American Classrooms, 1890–1990.* 2nd ed. New York: Teacher's College Press, 1993.

Dorn, Walter L. "The Debate over American Occupation Policy in Germany in 1944–1945." *Political Science Quarterly* 72, no. 4 (1957): 481–501.

Eisenberg, Carolyn. *Drawing the Line: The American Decision to Divide Germany.* Cambridge: Cambridge University Press, 1996.

Füssl, Karl-Heinz. *Die Umerziehung der Deutschen. Jugend und Schule unter den Siegermächten des Zweiten Weltkriegs 1945–1955.* Paderborn: Schöningh, 1995.

Füssl, Karl-Heinz and Gregory Wegner. "Education Under Radical Change: Educational Policy and the Youth Program of the United States in Postwar Germany." *History of Education Quarterly* 36, no. 1 (1996): 1–21.

Gaddis, John Lewis. *The United States and the Origins of the Cold War, 1941–1947.* New York: Columbia University Press, 1972.

Germany, 1947–1949: The Story in Documents. Washington, DC: Department of State, Office of Public Affairs, Publication 3556, European and British Commonwealth Series 9, 1950.

Gimbel, John. *A German Community Under American Occupation: Marburg, 1945–1952.* Stanford, CA: Stanford University Press, 1961.

Hall, Robert King. "The Battle of the Mind: American Educational Policy in Germany and Japan." *Columbia Journal of International Affairs* 2, no. 1 (1948): 59–70.

Hahn, H. J. *Education and Society in Germany.* New York: Berg, 1998.

Hearnden, Arthur. *Education in the Two Germanies.* Oxford: Basil Blackwell, 1974.

Hechinger, Fred M. "Germans Glibly 'Veto' American School Reforms." *The Washington Post,* September 28, 1947, B8.

Kellermann, Henry J. *Cultural Relations as an Instrument of US Foreign Policy: The Education Exchange Program between the United States and Germany, 1945–1954.* Washington, DC: U.S. Department of State Bureau of Educational and Cultural Affairs, 1978.

———. "Von Re-education zu Re-orientation: Das amerikanische Re-orientierungsprogramm im Nachkriegsdeutschland," in *Umerziehung und Wiederaufbau:Die Bildungspolitik der Besatzungsmächte in Deutschland und Österreich,* ed. Manfred Heinemann, 86–102. Stuttgart: Klett-Cotta, 1981.

Kliebard, Herbert M. *The Struggle for the American Curriculum, 1893–1958.* 2nd ed. New York: Routledge, 1995.

Knappen, Marshall. *And Call It Peace.* Chicago, IL: The University of Chicago Press, 1947.

Lange-Quassowski, Jutta-B. *Neuordnung oder Restauration? Das Demokratiekonzept der amerikanischen Besatzungsmacht und die politische Sozialisation der Westdeutschen.* Leske Verlag: Opladen, 1979.

———. "Amerikanische Westintegrationspolitik, Re-education und deutsche Schulpolitik," in *Umerziehung und Wiederaufbau: Die Bildungspolitik der Besatzungsmächte in Deutschland und Österreich.* ed. Manfred Heinemann, 53–67. Stuttgart: Klett-Cotta, 1981.

Occupation of Germany: Policy and Progress. Washington, DC: Department of State, Office of Public Affairs, Publication 2783, European Series 23, 1947.

Puaca, Brian M. *Learning Democracy: Education Reform in West Germany, 1945–1965.* New York: Berghahn Books, 2009.

128   *Charles Dorn and Brian Puaca*

Ravitch, Diane. *Left Back: A Century of Failed School Reform.* New York: Simon and Schuster, 2000.

"Report of the United States Education Mission to Germany." Washington, DC: Department of State, 1946.

Robinsohn, Saul and J. Caspar Kuhlmann. "Two Decades of Non-reform in West German Education," in *Education in Germany. Tradition and Reform in Historical Context,* ed. David Phillips, 311–330. New York: Routledge, 1995.

Schiller, Dietmar. "Schulalltag in der Nachkriegszeit," in *Schulreform—Kontinuitäten und Brücke. Das Versuchungsfeld Berlin-Neukölln,* vol. 2, ed. Gerd Radde and Werner Korthaase, et al., 29–40. Opladen: Leske & Budrich, 1993.

Shibata, Masako. *Japan and Germany under the US Occupation: A Comparative Analysis of Post-War Education Reform,* ed. Edward R. Beauchamp, *Studies of Modern Japan.* Lanham: Lexington Books, 2005.

Snell, John L. *Dilemma over Germany.* New Orleans, LA: Hauser Press, 1959.

Tent, James F. *Mission on the Rhine: Reeducation and Denazification in American-Occupied Germany.* Chicago, IL: University of Chicago Press, 1982.

Trachtenberg, Marc. *A Constructed Peace: The Making of the European Settlement, 1945–1963.* Princeton, NJ: Princeton University Press, 1999.

Tsuchimochi, Gary H. *Education Reform in Postwar Japan: The 1946 Education Mission.* Tokyo: University of Tokyo Press, 1993.

Tyack, David. *The One Best System: A History of American Urban Education.* Cambridge, MA: Harvard University Press, 1974.

White, Steven F. *Progressive Renaissance: America and the Reconstruction of Italian Education, 1943–1962.* New York: Garland, 1991.

Zilversmit, Arthur. *Changing Schools: Progressive Education Theory and Practice, 1930–1960.* Chicago, IL: University of Chicago Press, 1993.

Zink, Harold. *The United States in Germany, 1944–1955.* New York: D. Van Nostrand Company, 1957.

CHAPTER SIX

Demystifying the Divine State and Rewriting Cultural Identity in the U.S. Occupation of Japan

KENTARO OHKURA AND MASAKO SHIBATA

Religion in the Modernity Project

The purpose of this chapter is to discuss the implications of the U.S. postwar occupation of Japan for the political and educational positioning of religion in Japan. We posit that U.S.-sponsored educational reform after Japan's defeat in World War II marked a notable turning point in the treatment of religion within the Japanese state's modernization project. Moreover, the reforms significantly affected the ways in which the Japanese remember the war as part of their national history.

After World War I, the faithful in Christendom realized that "science devoid of ethics was deadly, and ethics divorced from faith was barren."[1] This lesson had not been learned at the end of *Pax Britanica*.[2] After World War II, the lesson became part of geopolitical policy in the age of American supremacy. The Western Allies believed that the ethical and spiritual revitalization of the defeated was essential for the reconstruction of defeated countries and the construction of a new peaceful world order. From the outset of the military occupation of Japan, one of the foremost aims of U.S. authorities was to purge the Japanese and their society of a militaristic and ultra-nationalistic mindset. The belief in the divinity of the emperor and the superiority of the Japanese race, which was based on the *Shinto* cult, was at the core of the moral condemnation of Japan by the United States. Ultimately,

Japan adopted the principle of the separation of state and religion in the Constitution of 1946 and in the Fundamental Law of Education of 1947, both drafted by the United States. Given the recent enforcement of, or debate about, the rewriting of these democratic canons for postwar Japan, this chapter revisits the genesis of the cultural significance of the demystification of the emperor, its relation to the modern Japanese nation as such, and the long-term implications of the U.S. occupation.

The other approach of this chapter is to scrutinize the knowledge of religion in Japanese education as a productive power. Religions as productive power embedded in systems of knowledge are different from those as a sovereign power, which struggles for the domination of the people. The notion of productive power enables one to see, feel, think, and act in effective ways while sovereign power resorts to coercion and repression against others. In the field of education, what is known about religion often normalizes children and adults, and eventually in the modern state, the nation as a whole. This occurs by informing a shared reasoning about what it means to be Japanese.

Ironic though it may seem, one can say that the search for a concept of religion is part of the modernity project. The conception of religion was an integral part of European secular traditions from the time of the Enlightenment. Thus, however inimical, modern secularity has religious roots itself.[3] While modernization is considered to be the process of rationalization, which is in principle, incompatible with religion, the modern economies and social configurations have been explained in terms of religion in the modern sociological analysis of religion on modernity.[4] Within the process of European modernization, reason has taken on the authority over public morality that was held by religion in an earlier period. This secularization paradigm, a model of public religious accommodation, itself grew out of Christianity.[5] Thus, in the European perception, modernity and religion are epistemologically a dichotomy. Moreover, in Europe, the departure from faith in the absolute and the value of plurality have been the major consequences and propositions of the rise of modernity.

In contrast, in Asia, notably in the eastern region, there has been a distinctive sense of pluralism in perceiving and positioning religion in society. In Japan, for instance, the dichotomy of *secular* and *religious* has only been introduced in line with the absorption of Western knowledge. In other words, the conceptualization of religion and its politico-intellectual application is an integral part of the modernization and Westernization project. Along with industrialization through advancing technology and the transfiguration of social structure, the conceptualization of religion is part and parcel of Japan's assimilation of the idea of the European civilization and a deliberate effort to be accepted as its member.[6] As the system of thought and as a powerful social institution, the Meiji leadership understood the significance of the concept of religion as a means of the politics of truth.[7] This is crucial in understanding the large number of Christians among the Japanese leadership, and above all, Japan's acceptance of the U.S. policies after her defeat in WWII.

It is generally said that the modern state is a secular society that removed the power of "church" over the collective will of the agents. Accordingly, the people are not subject to the intervention and the coercive power of church. However, they are continually governed by the knowledge of religion or what is reasoned about religion. Thus, the interest of this chapter also lies in the discursive governing of religion, not just the dogmatic effects of church. The knowledge of religion is central to the formation of the modern self and nation. Therefore, the denomination of church or religion becomes less problematic than the discursive power of religion in structuring attributes of the self in the modern state.

For example, in a similar manner of the nineteen-century's American "True Womanhood," the ideal of Japanese women was discussed in reference to Christianity as well as *Shinto*, a hybridized system of rites and worship to indigenous gods and Confucianism.[8] The Japanese in the early period of modernization learned values and styles of the West particularly through the adoption of Christianity. It was often believed that the transplantation of Christian values made Japan equivalent to the other, that is, "civilized" West.[9] Women in modern Japan were therefore educated to be "faithful, docile, enduring, and well-balanced."[10] Later, the value of "domesticity" was brought into the discussions over Japanese womanhood. These traits discursively disciplined actions of Japanese women, and particularly, "domesticity" produced practices of home management and economy for their specialized space. When the country addressed the modernization project, the way to view, reflect, and even discipline the self became similar to the countries of the West and in a sense involved a localization of Christianity.

By what is known about religion, rules are constituted for what it means for people to think and act "reasonably." The nation is thus imagined as a whole through not only political ideologies or capitalism, but it is tied together by each member sharing a secular knowledge of religion (i.e., searching for womanhood, or teaching morals), or having common faith in the nature of human beings (i.e., cultivating morality as discussed later), or through norms that produce moral relationships and social order among the people. It is important to point out that secular knowledge of religion fabricated in the school curriculum contributes to totalizing the people on the assumption that all the people can be virtuous in terms of "obedience to authority" or "cultivation of capability."

Mystification of the Emperor (1868–1945)

Since the restoration of the imperial throne in 1867, the sovereign power of the state has been endorsed to the emperor. Notwithstanding the somewhat random adoption of Western ideas initially, the German notions of the *Omnipotentz des Staates* captured supreme legitimacy in the formation of the *family state*. In it, the manipulation of *Shinto* played a central role in awakening nationalist sentiment among the public.

In the name of rationalization, the government also maneuvered the faiths and rituals of Shinto by mixing greatly varying local traditions into a simplistic linear theology. Shinto is not a monolithic set of practices throughout the nation. Paradoxically, this strong indoctrination of an unorthodox, polytheistic Shinto by the government actually alienated the people from religious institutions.[11] According to James Grayson, this "Shintoization" by the Japanese government made it difficult for the Japanese public to link their religious conduct with the acts of the military government based on the state Shinto ideology.[12]

In addition, the government's positioning of state Shinto in the public sector was manipulative. From 1889 on, the government recognized that state Shinto was the national ideology rather than a religion. Indeed, it was no longer a religion, but a political device for mobilizing the people for total war, and for hallowing self-sacrifice and death for the national deity. The heroic souls, or *eirei*, of the war-dead were enshrined by the government in the Yasukuni Shrine. The shrine, renamed Tokyo Shokonsha in 1879, was initially created in 1869 when the government enshrined 7,751 souls of dead soldiers who fought for the sake of the state and the emperor during the civil war in the Meiji Restoration. As of 2004, the shrine honored nearly 2.47 million heroic souls, including over 2.13 million from WWII.[13] Given its genesis, the Yasukuni Shrine has acted as the de facto national cemetery and has been recognized as such in most state affairs throughout its history, although no corpses were buried there. While safeguarding state Shinto as "nonreligious," the government attempted to exclude other religions, Christianity in particular, from the public sector in the name of the separation of state and religion. In this sense, the publication of *Conflict between Education and Religion* written by Imperial University professor, Inoue Tetsujiro, was one of the momentous incidents within which the governmental policy for religion and Christianity was made clear. In it, he overtly claimed that Christianity was antinational.[14]

Shushin as Secular Religion

In education, the government constructed state *Shinto* by systematizing nondogmatic *Shinto* theology and embedding Confucian ethics into it. In 1880, *shushin*, or moral education, became the leading subject within the curriculum of primary schools. The instillation of state Shinto now became legal. *Shushin* is "secular" namely because it is a mixture of Shinto, moralistic, and Confucian ethical maxims, which totalizes the peoples of different belief systems. Spiritual dogmas in religions were "sublimated" for the official curriculum and elaborated into a collection of moral stories by philosophers and educators. Thus, *shushin* does not merely represent the political system of imperialism, but it represents the national polity of the imperial state where the members of society are bound together as Japanese through the narrative of morals, a so-called *purified* knowledge that serves as a body of religious understandings. In 1890, the Imperial Rescript on Education was promulgated in the name of the emperor as the canon of public education. The recitation of the script and bowing to the imperial portrait were mandatory for *all* children.

Shushin appeared first in the curriculum as schools began to be centralized after the education edict of 1880. The textbooks of *shushin* were officially produced and revised to show what morals are practiced as the Japanese.[15] The content was organized to emphasize the learning of morals from great men and their practices. However, a nationalistic tone was not established in *shushin* in the initial stages, but gradually emerged over time. In the first edition, the prospectus encouraged teachers to focus on the motivation of students for learning morals. In other words, the indoctrination of specific moral lessons was strictly deterred. However, there were some exceptions in lessons such as respecting women, being aware of the system of the self-governing nation-state, understanding public health, and civic mindedness. These new concepts produced a distinction and differentiation of practices, particularly in the spheres of gender and society. The disciplines and rules were inscribed in the textbook and embodied what men/women work at, how one can act in public/private, and what the Japanese/other nations can do. For example, the first edition of the *shushin* textbook recognizes individuals as free and autonomous, and at the same time, seeks to "save" the souls of children from the feudalistic as well as savage condition of the country at the time.

The moral textbooks in the following editions changed tone in a dramatic way. They began to stress the values of the imperial tradition of Japan including admiring the (imperial) ancestry and history, serving the emperor as well as master, and joining in the military service. Less emphasis was placed on the motivation of students as the agentive individual. The Imperial Rescript on Education was also issued as the supreme law of education. The values imparted on the child included the idea of seeing oneself as part of a hierarchical order; thus, the child learned to admire older relatives, parents, seniors, and ultimately the emperor. Practices framed by the seniority system of imperialism were praised and encouraged in *shushin*. This hierarchical seniority system contributed to unifying and maintaining relationships among citizens of the nation-state.

The textbook was subsequently revised a total of four times before the imperial Japan turned into a complete totalitarian state in the 1940s. All of the textbook editions were composed of moral content recognizable along with the grade of student. Motivation was given as an effective means for the self-fulfillment of an individual who was about obtaining moral values. This method, called the Herbartian method, was adopted from Germany and contributed to teaching morals in an a priori manner.

The elements of mythology were continuously embraced in the fifth edition, but the sacredness of the emperor was more accentuated than ever before. *Shushin* was edited for consistency with other school subjects including history, geography, and language. *Shushin* was interconnected to other subjects to endorse its authenticity profoundly rooted from the legend, motherland, and vernacularism of the imperial Japan.[16] The body knowledge of the fifth edition hence embodied the imperial system of national integration and the homogenization of the people, but it did not encourage a social system of civic participation.

Under the totalitarian regime, the *shushin* textbook began its lesson with teaching that the ancient gods were the imperial ancestry having created the land of Japan.[17] Japan was repeatedly described as "a land of the gods." It was then righteous for the children to dedicate themselves to the emperor as the absolute. Serving the country is equivalent of devoting oneself to the gods.[18] The souls of all the war dead are honored at the Yasukuni Shrine, where the emperor and empress customarily made visits for pacifying the souls of the devotees. In short, this edition symbolically described moral virtues of "loyalty" and "filial piety" as necessary for family ties (i.e., familism) and as nationalism underpinning a special nationhood centered on the emperor.[19]

Shushin functioned as a narrative of religion with which the people identify themselves through the emperor as a descendent of the creator. *Shushin* also functioned as a narrative of the land enfranchised by Him. But, there are few precepts or tenets translated from the words of the emperor and explained in the narrative. As a secular knowledge of Shinto religion, *shushin* demanded students to follow particular manners (i.e., loyalty and filial piety) in which the imperial subjects had historically constructed the relationship with the descendents in reference to the mythology and traditional rituals of the imperial household. Shinto itself was not theologically conceptualized as is often observed in the Western religions, but it included religious sentiments for the sake of Japan's totalitarianism and modernization project.[20]

Reshaping of the Modernity Project

After Japan's defeat in WWII, the country had been under the military occupation of the Allied Powers, in effect under the United States. The occupation lasted until 1952, when the San Francisco Peace Treaty, or the Treaty of Peace with Japan, came into force and the recovery of Japan's sovereignty was verified by 49 countries chosen by the Allies. Among social institutions, one of the first reforms that the U.S. military government launched was that regarding religion. The whole purpose of this reform was the elimination of state Shinto (i.e., the perceived basis of Japan's ultra-nationalism and militarism). On December 15, 1945, four months after Japan's surrender, the U.S. authorities issued a military order to "separate religion from the State, to prevent misuse of religion for political ends, and to put all religions, faiths, and creeds upon exactly the same legal basis, entitled to precisely the same opportunities and protection."[21] This order, which prohibited the "sponsorship, support, perpetuation, control, and dissemination of Shinto by the Japanese national, prefectural, and local governments, or by public officials, subordinates, and employees acting in their official capacity," was called in short the Shinto Directive. On the eve of its announcement, the U.S. staff in the information and education section thought that "all hell would break loose."[22]

In occupied Japan, religious authorities and the occupation authorities got along relatively well. Notably, in the initial stage, the Americans paid close attention to Christian activities in wartime Japan. Earlier than the Shinto Directive, the military government ordered the Japanese government to dismiss teachers and administrators from Christian schools due to their wartime acts of "inexcusable and unjustifiable subversion... to militaristic and ultra-nationalistic ends."[23] This directive was delivered to about 100 Christian educational institutions throughout the nation, and over 80 institutions were screened within a few days.[24] Indeed, the principle idea of *religious freedom* in the United States is applied for safeguarding and spreading Christianity, as stated in the Act of Toleration passed in the Maryland colony in the 1640s.[25] This idea was certainly valid in the middle of the twentieth century. At home in the United States, while sensitive critics recognized the violation of the state-church separation by their own army in Japan, some *naive* people said that "If you are not helping Christianity, what are you doing?"[26]

After parting from state sponsorship, most Shinto shrines came under the umbrella of one of the religious corporations, the Association of Shinto Shrines or *Jinja Honcho*. Among the member shrines of the association, the Grand Shrine of Ise resides in its apex as the most honored place in spiritual and historical terms. The association is the largest and the most politically powerful religious corporation in Japan and controls about 80,000 Shinto shrines as of 2006 throughout the country. But a number of the traditional and influential shrines, including Yasukuni, are not members of this association.

Regarding the treatment of the emperor, the U.S. State Department and the Far Eastern Commission of the Allies decided wisely, in the American eyes, to allow Japan to retain the imperial household system and the emperor himself, but not his divinity. In December 1945, a crucial suggestion was delivered to the State-War-Navy Coordinating Subcommittee for the Far East from Edwin Reischauer: "Any attempt to persuade the emperor to participate in his own 'debunking' should be made in such a manner as to be unknown to the Japanese people and should be handled with such diplomacy as to give no suggestion of compulsion."[27] On January 1, 1946, the emperor made a "voluntary" refutation of the myth of state Shinto by pronouncing that "[People] are not predicated on the false conception that the Emperor is divine, and that the Japanese people are superior to other races and destined to rule the world."[28]

State sovereignty shifted to the people from the emperor, but he remained on the throne as "the symbol of the State and of the unity of the people," according to the newly U.S.-drafted new constitution. Notwithstanding strong objection voiced by the other Allies and inter alia Japanese communists, the U.S. authority also decided to exempt the emperor from indictment as a war criminal at the International Military Tribunal for the Far East in Tokyo or the Tokyo Trial (May 3, 1946–November 12, 1948). The United States seems to have made a "wise" decision indeed, considering the significance of the verdicts given at this trial throughout the postwar period. On the whole, the United States avoided the Japanese harboring of lasting resentment toward the Americans and above all the "communization" of East Asia, especially in the apparent decline of Chiang Kai-shek in China. Indeed, part of the Japanese leadership and the public felt that it was fortunate for Japan that the country was occupied by the Americans and not by the other Allied Powers.[29]

In education, the American education mission to Japan in 1946 critically pointed out that Shinto as the state religion (or cult) prevented the dissemination of democracy and liberalism.[30] The group added to the new meaning of religion in Japan:

> Some democracies separate church and state through fundamental law. They do so in order that both religion and government may contribute their utmost to a complete life. We believe that democracy properly conceived has this in common with religion, for in quest of the spiritual life, it emphasizes the dignity and worth of individual human beings, together with characteristics of brotherhood common to all. Liberty carried to abnormal lengths will yield irresponsibility, anarchy, and chaos. But, equally fatal to the human spirit can be a oneness that reduces the individual to a meaningless part of a stifling whole.[31]

It is herein underlined that the worth of religion is not necessarily in conflict with that of the individual.[32] In other words, the separation of church and state became problematic if religion facilitated the homogenization of the people and forbade the natural development of the individual. "Individuality" was thus conceptualized and produced as the major concern for new education and schooling after the separation of church and state.

Both the Fundamental Law of Education and the School Education Law of 1947 subsequently reflected the mission report. The former established the aims of education as the "full development of personality" and "esteem[ing] individual value," and the later placed that of schooling as "education according to the development of [children's] minds and body."[33] These ideals are a great contrast to that of the prewar education system and its totalitarianism centered upon the emperor. More importantly, this belief in individuality required the use of psychology in implementing moral education.

Moreover, instruction in *shushin* was suspended and has never been restored in the school curricula. Although *shushin* was eventually replaced by *doutoku* in the 1950s, social studies was temporally brought into schools and substituted for teaching morals in society and structuring a collective identity. Social studies thus embodied the spirit of civil society, and more importantly, the identity of cosmopolitanism where the people may be associated through *cultural aspects* of religions.[34] Teachers did not directly teach Buddhism, Christianity, or Shintoism, but took students to temples, churches, and even shrines to study them as part of common legacy. In short, both Americans and Japanese, though with different goals in mind, appreciated the significance of religion for building the nation unified. Within these religious transfigurations, despite a number of the American intrusions into ecclesiastical issues, there was no acute *Kulturkampf* between religious organizations and the military government in Japan as was witnessed in occupied Germany.

Doutoku as Faith

In postwar discussions on moral education, *doutoku* came up as the prime scheme. The Ministry of Education and leading educators began to discuss if schools should prepare a single school subject for moral education instead of social studies.[35] Introduced as a school subject in 1958, *doutoku* received a heavy influence from the study of developmental psychology. *Doutoku* was established with a new educational aim: *cultivating morality* in the mind of students, but not merely through inculcating moral principles and values.[36] This way of viewing students required the use of psychology conceptualized by the individual potentiality and growth in the classroom. *Cultivating morality* was crucial for those Japanese who were about to live in a democratic and international (or *cosmopolitan*) society since students as citizens of the future need not only to know what is right or wrong, but also as agentive individuals they need to "possess the mind of morality which enables ones to make his judgment (on good or bad)...by stimulating [their] interests and concerns as well as providing various experiences."[37] The child as the agentive individual is produced by developmental psychology, which constructs the child as a problem-solver, active learner, and constructive member of the community.[38]

The application of developmental psychology to *doutoku* meant not merely the denial of *shushin* or the introduction of a scientific approach to pedagogy. It made possible to give recognition to the concepts of "nature," "will," and moral judgment of the individual, which are all part of Christian beliefs and concepts of divinity.[39] In referring to Kant's, Rousseau's, and Pestalozzi's views on humanity and individuality, the postwar educators of Japan understood the philosophy of developmental psychology, and realized that this view is different from the child who is guided only by the truth as revealed by God.[40] In consequence, only when the nature in a child is subject to cultivation and development can the morality in the mind be fully realized.

In the discussions on the instalment of *doutoku* as a school subject, what should be taught in *doutoku* remained unanswered. But, the discussants agreed upon an approach that fostered in the mind and attitudes of student a continuous "searching for truths." The teachers' manual in 1958 explained the moral mind as endowed needs, which are to be developed along with social needs. The moral mind is potentially capable of being aware of what is good and bad in society.[41] Thus, school subjects were organized to correspond as students' experiences expanded from the place of family to the community to which they belong. To put it differently, the course was to facilitate students "to ask themselves" or to search for "truths" from what they see, feel, and consider everyday (i.e., *internalization* and *identification* in psychological terms). In moral education, the value of life cannot be taught without the experience that the student was loved by somebody in the family or a neighbor, classmate, and community. This type of pedagogical approach was to ensure the agency of child, epistemologically speaking.

When virtue was not subject to indoctrination over child, but to help the child "recall" in Plato's conception, teachers directed their gaze to the innate nature of child. This way of looking at the child regards human nature as fundamentally good and recognizes the power of psychology as an avenue that appropriately develops the nature of the child. In turn, this nature is then seen as the site of control for moral education, signifying desire, will, and self-government.

The Revival of Shinto in the National Memory

While state Shinto was eliminated and the principle of state-religion separation was introduced by the U.S. occupation, the relation of the state to the Yasukuni Shrine was maintained for the nation in remembering the war in its own terms. As the ultimate settlement of WWII, the Allies and the Japanese government signed the San Francisco Peace Treaty. By this, as mentioned earlier, Japan's sovereignty was restored. In turn, Japan recognized the independence of her former colonies, including Korea and Formosa, and renounced all special rights benefited from 1901 in mainland China. At the same time, Japan accepted the judgments of the Tokyo Trial, and agreed that the reduction of the sentences of the war criminals or their parole was only possible in accordance with the Allied Powers consent related to each sentence. All these resolutions were made in the absence of Korea and China, based on the Allies' decision not to include these nations.

In contrast to the case of Germany, Japan's acts of atrocities conducted during WWII were treated as acts of the elite and not that of the whole Japanese populace. The relative simplicity of the screening of Japanese militarists versus the size and complexity of that of the Germans demonstrated this American perception.[42] Therefore, for the victims of Japan's wartime invasion in particular, the treatment of the war criminal was politically and diplomatically significant as a key question in resolving Japan's responsibility for the war.

However, eventually some of the war criminals began to be released from prison after the decision of the National Diet in June 1952. Of these released criminals, some regained their pre-war leading position in the State affairs. From the 1950s, the Yasukuni Shrine began to enshrine B and C Class war criminals with the support of the Ministry of Welfare (currently the Ministry of Health, Labour and Welfare). Eventually both parties decided to confer the same treatment for the highest-ranking war criminals, A Class.[43] This was also about the time that the country had enjoyed the economic growth higher than any period since the end of the war, and thus, the people gradually overcame the malaise of wartime defeat. The Central Council commissioned by the Ministry of Education championed educational reforms that retained "true Japanese" identity based on the cultural legacy of Japan and the Emperor.[44] For example, they recommended that schools restore "religious sentiments" in moral education and suggested that students value something sacred through which "the lives of parents, the races and eventually all human beings are generated."[45]

Thus, Yasukuni's treatment of the war criminals and the perpetuation of the shrine represent the government's view of Japan's conduct in WWII. The government's spotlighting the Yasukuni Shrine demonstrates its historical manipulation and its deliberate intention to legitimize nationhood.[46] The core of the Yasukuni Issue is its role of acting as an iconic symbol of WWII, framing the war as a sad, but not a bad, morally wrong war for Japanese citizens. The shrine's role in this sense was important, especially after the nullification of the nationalism, which was based on imperial mysticism. Yasukuni's view on the war is unacceptable by those who suffered wartime acts of atrocity at the hand of the Japanese military. The Yushukan museum, built within the shrine's site, exhibits the weapons of the military and a short movie, which actually justifies Japan's invasion into the Asian continent repeatedly plays. In the analysis of the role and function of Yasukuni's rituals, healing the wounds and sorrow of the family members of the war dead often becomes central. They often suffer from a sense of guilt for having survived the war and making their livelihood on a state stipendiary at the sacrifice of their sons. In this sense, some of the family members have sought a healing sanctuary and nostalgic aura within the Yasukuni Shrine as well as its argument, in which the price of the war is sanctioned and the war itself is legitimized.

Conclusion

A concluding question can be posed: what in essence did the U.S. occupation mean for the Japanese? The U.S. occupation and its aftermath have exposed to particular aspects of Japan's modernity project in terms of the state-religion relationship, a notion of faith, and a particular conception of *progress* and a particular perception of the nation as a collective body. At the point of the collapse of the Japanese emperor state, its modernity was regarded as an unfinished project. In this sense, the mission of the U.S. authorities was seen as a process of completing Japan's modernity project. Japan's modernity project would continue within a search for a specific notion of *progress*: making Japan fully modern.

In concrete, despite the introduction of the principle of state-religion separation, the Yasukuni Shrine survived as a powerful political and social institution. By acting as such, Yasukuni has been used by the government for its political ends. At the same time, Yasukuni has maintained an indissoluble relationship with the Japanese people through the shared experience of the *sad* war. As it was planned in the last third of the nineteenth century, the Shinto shrine has remained as an important iconic symbol of the nation and the national history. In other words, it has been articulating and re-articulating the narrative of modern Japanese nationhood. Furthermore, "truth" is not just told to be what the God reveals, but what is drawn from human nature, without the intermediation of the state.

The initial idea of the Allied occupation was to create a basis for the construction of a peaceful world order after WWII by deconstructing and reconstructing Japan and other Axis countries. Moreover, the idea of the U.S. occupation in Japan was to invite Japan to be a member of the postwar world community. However, in the process of completing the modernization project, Japan invented a unique rhetoric which held that Japan could exist as a real international member of society only if the Japanese rediscovered a genuine identity with a legacy that included the imperial household and the emperor.[47] The binary opposition between the Japanese self and foreign others came to be maintained as a framework to talk, act, and think about the postwar Japanese in many places.

The framework of modernity supposedly completed with the presence of the occupation enables one to recognize new problems in the postwar Japan. The end of the war merely changed the way in which the Japanese discovered the self in relation to others.

Notes

1. "The Moot Papers" (Special Collection Archives of the Institute of Education, University of London, 1939–1942). The Moot was a Christian discussion group, related to the Christian Frontier Council, and made up of eminent philosophers for the reconstruction of German education after WWII.
2. Niall Ferguson, *The War of the World: History's Age of Hatred* (London: Allen Lane, 2006).
3. Peter Berger, *The Sacred Canopy: Elements of a Sociological Theory of Religion* (New York: Doubleday, 1966).
4. Émile Durkheim, *The Elementary Forms of the Religious Life* (Oxford: Oxford University Press, 2001 [1915]); Max Weber, *The Protestant Ethic and the Spirit of Capitalism* (London: George Allen & Unwil Ltd. 1976 [1904–05]).
5. Elizabeth S. Hurd, "The Political Authority of Secularism in International Relations," *European Journal of International Relations* 10, no. 2 (2004): 255.
6. Dana Buntrock, "Without Modernity: Japan's Challenging Modernization," *Architronic* 5, no. 3. (1996): 1–5. Jun'ichi Isomae, *Kindai Nihon no Shuukyou Gensetsu* (Tokyo: Iwanami Shoten, 2003).
7. Aziz Talbani, "Pedagogy, Power, and Discourse: Transformation of Islamic Education," *Comparative Education Review* 40, no. 1 (1996): 66–82.
8. On America's True Womanhood, I refer to the discussions with Barbara Welter, "The Cult of True Womanhood: 1820–1860," *American Quarterly* 18, no. 2 (1966): 151–174; Tracy Fessenden, "Gendering Religion," *Journal of Women's History* 14, no. 1 (2002): 163–169.
9. Herein I refer to the discussions with Tsuda Umeko, Oe Sumi, and Shimoda Utako.
10. Monbusho, "Koto Jyogakko Kitei," in *Gakusei 100nenshi: Kijyutsu hen-Shiryo hen*, ed. Monbusho (Tokyo: Teikoku Chiho Gyosei Gakkai, 1895). Kazuko Nagahara, "Ryosai Kenbo Shugi ni Okeru ie to Shokugyo," in *Nihon Jyosei-shi vol. 4* (Tokyo: Daigaku Shuppankai, 1985).

11. Yuki, Shiose, "Japanese Paradox: Secular State, Religious Society," *Social Compass* 47, no. 3 (2000): 317.
12. James Huntley Grayson, "'Shinto' and Japanese Popular Religion: Case Studies of Multi-variant Practice from Kyushu and Okinawa," *Japan Forum* 17, no. 3 (2005): 347–367.
13. Jinja Yasukuni, www.yasukuni.or.jp/annai/gaiyou.html (accessed March 15, 2006).
14. Tetsujirou Inoue, *Kyoiku to shukyo no shototsu* (Tokyo: Keigyosha. 1893).
15. Monbusho, "Dai'ichiji Kokutei Shogaku Shushinsho Hensan Shuihoukoku," in *Doutoku Kyoiku Shiryo Shusei vol. 2.* ed. Miyata Takeo (Tokyo: Dai'ichi Houki, 1904), 69.
16. Ibid., 932.
17. bid., 1004.
18. Ibid., 1026.
19. Klaus Antoni, "Yasukuni-Jinjya and Folk Religion: The Problem of Vengeful Spirits," *Asian Folklore Studies* 47 (1988): 123–136.
20. Shigeyoshi Murakami, *Kokka Shinto* (Tokyo: Iwanami, 1970).
21. SCAP, "Abolition of Governmental Sponsorship, Support, Perpetuation, Control, and Dissemination of State Shinto (*Kokka Shinto, Jinja Shinto*)" of December 15, 1945, Memorandum of SCAP, CI&E for Imperial Japanese Government, (1945b).
22. William Woodard, *The Allied Occupation of Japan 1945–1952 and Japanese Religions* (Leiden: E. J. Brill. 1972), 69.
23. SCAP, "Violation of Religious Freedom" of October 24, 1945, Memorandum of the Office of Supreme Commander of the Allied Powers for imperial Japanese Government, (1945a).
24. Reiko Yamamoto, "Kyoshoku tsuiho," in *Sengo kyoiku kaikaku tsushi*, ed. Meisei University Sengo Kyoiku-shi Kenkyu Centre (Tokyo: Meisei University Press, 1993), 235–236.
25. R. Audi, "The Separation of Church and State and the Obligations of Citizenship," *Philosophy and Public Affairs* 18 (1989): 259–296.
26. Woodard, *The Allied Occupation*, xi.
27. SWNCS (State-War-Navy Coordinating Subcommittee for the Far East), "Treatment of the Institution of the Emperor" of December 11, 1945, "Post World War II Foreign Policy Planning: State Department Records of Harley A. Notter, 1939–1945," 1945, #1520 H-128.
28. GHQ, *Education in the New Japan vol. 1* (Tokyo: Civil Information Education Section, Education Division, 1948), 77–78.
29. Yoshishige Abe, *Sengo no jijoden* (Tokyo: Shinchosha, 1959), 59.
30. Report of the United States Education Mission to Japan: Submitted to the Supreme Commander for the Allied Powers, Tokyo, March 30, 1946. S. l.: s. n. Intro.
31. Ibid., 9.
32. For more details on the relationship between the Constitution and the Education Law, please refer to Matsuoka Shigeo, "Gendai Shukyo Kyoiku niokeru Kuniku no Mondai," *Kyouikugaku Kenkyu* 47, no. 2 (1980): 117–125.
33. GHQ, *Education in New Japan vol. II* (Tokyo: Civil Information Education Section, Education Division, 1948), 109–119.
34. Kyouiku Sasshin I'inkai, *Sengo Kyoiku no nakano Doutoku Shukyo*, ed. Kaizuka Shigeki (Tokyo: Bunka Shobo Hakubunsha, 1948).
35. The *Kyoiku katei shingikai* of 1959 is the discussion group referred to here.
36. Monbusho, "Shogakko Doutoku Shidosho," in *Doutoku Kyoiku Shiryo Shusei vol. 3*, ed. Miyata Takeo, (Tokyo: Dai'ichi Houki, 1958).
37. Ibid., 655.
38. Thomas S. Popkewitz, "The Reason of Reason. Cosmopolitanism and the Governing of Schooling," in *Dangerous Coagulations? The Uses of Foucault in the Study of Education*, ed. Bernadette Baker and Katharina Heyning (New York: Peter Lang, 2004).
39. Ibid., 194–195.

144   Kentaro Ohkura and Masako Shibata

40. Takashi Ota, "Gendai Shakai to Kodomo no Hattatsu," in *Kodomono Hattatsu to Kyouiku vol. 1* (Tokyo: Iwanami shoten. 1979).
41. Monbusho, "Shogakko Doutoku Shidosho," 669.
42. Masako Shibata, "Doitsu 'Sai-Kyoiku' to Shukyo Kyoiku: 1933 *Reichkonkordat* Shori-Mondai wo Meguru America no Tai-Doku Senryo-Seisaku," *Sengo Kyoiku-shi Kenkyu* 17 (2003): 37–49.
43. Asahi Shimbin: 2007.
44. Chukyoshin, "Kitaisareru Ningenzo," in *Sengo Kyoiku no nakako Doutoku Shukyo*, ed. Kaizuka Shigeki (Tokyo: Bunka Shobo Hakubunsha, 1966).
45. Ibid., 234.
46. Hardacre: 1989.
47. J. Victor Koschmann, Intellectuals and Politics, in *Postwar Japan as History*, ed. Andrew Gordon (Berkeley, CA: University of California Press, 1993).

Bibliography

Abe, Yoshishige. *Sengo no jijoden*. Tokyo: Shinchosha, 1959.
Amano, Teiyu. Kokumin jissen yoryo. *Sengo Kyoiku no nakano Doutoku Shukyo*, ed. Kaizuka Shigeki. Tokyo: Bunka Shobo Hakubunsha, 1953.
Antoni, Klaus. Yasukuni-Jinjya and Folk Religion: the Problem of Vengeful Spirits. *Asian Folklore Studies* 47 (1988): 123–136.
Audi, R. The Separation of Church and State and the Obligations of Citizenship. *Philosophy and Public Affairs* 18 (1989): 259–296.
Berger, Peter. *The Sacred Canopy: Elements of a Sociological Theory of Religion*. New York: Doubleday, 1966.
Buntrock, Dana. Without Modernity: Japan's Challenging Modernization. *Architronic* 5, no. 3 (1996): 1–5.
Chukyoshin. Kitaisareru Ningenzo, in *Sengo Kyoiku no nakako Doutoku Shukyo*, ed. Kaizuka Shigeki. Tokyo: Bunka Shobo Hakubunsha, 1966. Durkheim, Émile. *The Elementary Forms of the Religious Life*. Oxford: Oxford University Press, 2001 [1915].
Ferguson, Niall. *The War of the World: History's Age of Hatred*. London: Allen Lane, 2006.
Fessenden, Tracy. Gendering Religion. *Journal of Women's History* 14, no. 1 (2002): 163–169.
GHQ (General Headquarters, Supreme Commander for the Allied Powers). *Education in the New Japan vol. I*. Tokyo: Civil Information Education Section, Education Division, 1948a.
———. *Education in New Japan vol. II*. Tokyo: Civil Information Education Section, Education Division, 1948b.
Grayson, James Huntley. "Shinto" and Japanese Popular Religion: Case Studies of Multi-variant Practice from Kyushu and Okinawa. *Japan Forum* 17, no. 3 (2005): 347–367.
Hurd, Elizabeth S. The Political Authority of Secularism in International Relations. *European Journal of International Relations* 10, no. 2 (2004): 235–262.
Inoue, Tetsujirou. *Kyoiku to shukyo no shototsu*. Tokyo: Keigyosha, 1893.
Isomae, Jun'ichi. *Kindai Nihon no Shuukyou Gensetsu*. Tokyo: Iwanami Shoten, 2003.
Kaizuka, Shigeki. *Sengo Kyoiku no nakano Doutoku Shukyo*. Tokyo: Bunka Shobo Hakubunsha, 2003.
Koschmann, Victor J. Intellectuals and Politics, in *Postwar Japan as History*, ed. Andrew Gordon, 395–423. Los Angeles and Berkeley, CA: University of California Press, 1993.
Kyouiku Sasshin I'inkai. *Sengo Kyoiku no nakano Doutoku Shukyo*, ed. Kaizuka Shigeki. Tokyo: Bunka Shobo Hakubunsha, 1948.

Monbusho. Koto Jyogakko Kitei, in *Gakusei 100nenshi: Kijyutsu hen-Shiryo hen,* ed. Monbusho. Tokyo: Teikoku Chiho Gyosei gakkai, 1895.

———. Dai'ichiji Kokutei Shogaku Shushinsho Hensan Shuihoukoku, in *Doutoku Kyoiku Shiryo Shusei* vol. 2, ed. Miyata Takeo. Tokyo: Dai'ichi Houki, 1904.

———. Shogakko Doutoku Shidosho, in *Doutoku Kyoiku Shiryo Shusei vol. 3,* ed. Miyata Takeo. Tokyo: Dai'ichi Houki, 1958.

———. Shogakko Shidosho Doutoku-hen, in *Monbusho Gakushu Shidosho vol. 27,* ed. Namura Kikuji. Tokyo: Osorasha, 1969.

The Moot. "The Moot Papers" (Special Collection Archives of the Institute of Education, University of London). The Moot was a Christian discussion group, related to the Christian Frontier Council, and made up of eminent philosophers for the reconstruction of German education after WWII, 1939–42.

Murakami, Shigeyoshi. *Kokka Shinto.* Tokyo: Iwanami, 1970.

Nagahara, Kazuko. Ryosai kenbo shugi ni okeru ie to shokugyo, in *Nihon Jyosei-shi vol. 4.* Tokyo: Daigaku Shuppankai, 1985.

Ota, Takashi. Gendai shakai to kodomo no hattatsu, in *Kodomono Hattatsu to Kyouiku vol. 1.* Tokyo: Iwanami shoten, 1979.

Popkewitz. Thomas S. The Reason of Reason. Cosmopolitanism and the Governing of Schooling, in *Dangerous Coagulations?* ed. Bernadette Baker and Katy Heyning, 189–223. New York: Peter Lang, 2004.

Sasagawa, trans. *GHQ nihon senryoushi.* Tokyo: Nihon Tosho Senta, 2000.

SCAP. "Violation of Religious Freedom" of October 24, 1945, Memorandum of the Office of Supreme Commander of the Allied Powers for Imperial Japanese Government, 1945a.

———. "Abolition of Governmental Sponsorship, Support, Perpetuation, Control, and Dissemination of State *Shinto (Kokka Shinto, Jinja Shinto)*" of December 15, 1945, Memorandum of SCAP, CI&E for Imperial Japanese Government, 1945b.

Shibata, Masako. Doitsu "sai-kyoiku" to shukyo kyoiku: 1933 *Reichkonkordat* shori-mondai wo meguru America no tai-doku senryo-seisaku. *Sengo Kyoiku-shi Kenkyu* 17 (2003): 37–49.

———. *Japan and Germany Under the U.S. Occupation: A Comparative Analysis of Post-War Education Reform.* Lanham: Lexington Books, 2005.

Shiose, Yuki. Japanese Paradox: Secular State, Religious Society. *Social Compass* 47, no. 3 (2000): 317.

SWNCS (State-War-Navy Coordinating Subcommittee for the Far East). "Treatment of the Institution of the Emperor:" of December 11, 1945, in *"Post World War II Foreign Policy Planning: State Department Records of Harley A. Notter, 1939–1945,"* #1520 H-128, 1945.

Talbani, Aziz. Pedagogy, Power, and Discourse: Transformation of Islamic Education. *Comparative Education Review (Special Issue on Religion)* 40, no. 1 (1996): 66–82.

U.S. Education Mission. Report of the United States Education Mission to Japan: Submitted to the Supreme Commander for the Allied Powers, Tokyo, March 30, 1946. S. l.: s. n, 1946.

Weber, Max. *The Protestant Ethic and the Spirit of Capitalism.* London: George Allen & Unwil Ltd, 1976 [1904–05].

Welter, Barbara. The Cult of True Womanhood: 1820–1860. *American Quarterly* 18, no. 2 (Summer, 1966): 151–174.

Woodard, William. *The Allied Occupation of Japan 1945–1952 and Japanese Religions.* Leiden: E. J. Brill, 1972.

Yamamoto, Reiko. Kyoshoku tsuiho, in *Sengo kyoiku kaikaku tsushi,* ed. Meisei University Sengo Kyoiku-shi Kenkyu Centre, 235–236. Tokyo: Meisei University Press, 1993.

Yasukuni Jinja. 2006. www.yasukuni.or.jp/annai/gaiyou.html (accessed March 15, 2006).

CHAPTER SEVEN

German Postwar Educational Reform and the "American Way of Life"

THOMAS KOINZER

In the spring of 1963 one of many questions by German educators in preparing their summary conference at the American Jewish Committee's headquarters in New York City after a study trip through the United States was: "Do Americans think...that their education is...worth being copied?"[1] The question examined how a specific way of life in a modern society influences education. It was concerned with the role of schools as social institutions with a specific (democratic) style of teaching and learning that prepared adolescents for life in a democratic society and that taught them to withstand chauvinistic and nationalistic ideas. Almost 20 years after the end of World War II, West Germany was seen by many Americans and some Germans as a fragile democracy. After almost 20 years of postwar economic and social growth in West Germany, after the country had become a member of NATO and many other international organizations, after democratic political structures and institutions had been in place for almost two decades, a "democratic spirit" among its citizens and especially among the young was still perceived as absent. Furthermore, following American reeducation and reorientation approaches[2] in the years after the war, the educational sector was still viewed as especially deficient. The school system was characterized by John Slawson, vice president of the American Jewish Committee, after his visit to West Germany in 1959 as "a tradition-bound, authoritarian school system, an emotionless, subject matter-obsessed system of lecture and rote—a factory in which children were deprived of any opportunity to think or act democratically."[3] For Americans, the structured school model in West Germany and the perceived rigidity of teaching styles were responsible for authoritarian, chauvinistic, and nationalistic attitudes that led to the rise of National Socialism and were still widespread many years after the war.

American educational interventions in West Germany took on many different forms. One such form was the institution of German-American exchange programs for people from all walks of life, especially students and professionals in the field of education. Germans had visited the United States and studied the "American Way of Life" as it played out in education since the early twentieth century.[4] After World War II, however, those visits were part of an educational effort to civilize, demilitarize, and democratize Germany. After the German militarism and Nazism had subdued Germans, education would now eradicate widespread nationalism and chauvinism.[5] "Exchange" more or less meant bringing Germans to the United States to introduce them to the "American way of life" and to show them a democracy in action. It was hoped that, upon their return, they would apply American ideas to the democratization of West Germany and of its central institutions, especially those in the educational and school sector. The exchange and travel programs, initiated and organized by American military and civil authorities and later facilitated by nongovernmental American and German organizations and foundations, were among the most lasting initiatives to democratize West Germany and its educational and school sector. This chapter focuses on the exchanges that took place after World War II in order to introduce West Germans to the American way of life and to confront them with a style of (every day) living that would thoroughly immunize them against chauvinism, nationalism, and fascism. This chapter especially concentrates on the crucial role education and schooling played in those exchanges.

After a short introduction to the exchange programs, including their aims and their size, I will focus on a specific travel program for educators, teachers, school administrators, educational politicians, university lecturers, and professors that were introduced in a phase of a second, further American reeducation or reorientation effort in the 1960s known as *German educators missions*. Organized by the American Jewish Committee (AJC) and the German philosopher and sociologist Max Horkheimer in association with Theodor W. Adorno, this program sought to provide opportunities to learn about the American way of life in education, and to initiate and foster German initiatives to reform the educational and school system in West Germany. The German visitors were supposed to introduce the ideas they had encountered and the specifically American way of life in education into the German discourse, thereby putting pressure on the German educational reform process, the discourse, and after all, the educational and school practice.

The article portrays the German educators missions during the 1960s and early 1970s, a period that is commonly known as the *German education reform era* and characterized by a rapid and fundamental change in the educational system and its institutions.[6] The study trips to U.S. schools and classrooms and the observations of the American way of life in education by German educators were described against the background of an image of the "German school" and school system in dire need of democratic reform. This chapter will make the argument that the "American way" in education, the American philosophy of education, and the educational practice as perceived by the German visitors influenced the German educational and school reform process and discourse. In that perspective, the perceived American way of life in education had an externalization effect and functioned as an "additional meaning"[7] to support specific German positions on educational and school reform. However, the American way of life in education had also a negative image and was a threat to the traditional German concept of institutionalized education and the concept of *Bildung*. The American way of life in education was therefore used as a reference by both proponents and opponents of school reform in order to bolster their respective positions. In a "twin reception" (or "doubled-bound transfer"), this concept entered educational discourse on the means and ends of schooling and school practices in a democratized, modern Germany.

Exchange Programs and the "American Way of Life" in Education—the Early Years

As Henry J. Kellermann, a staff person of the American Foreign Office who had emigrated from Germany, noted, the exchange with Germany was the "largest and most ambitious program of cultural exchanges ever undertaken by any government as an instrument of foreign policy."[8] The purpose of that kind of foreign policy was to introduce the American way of life through a very personal and physical encounter. As Kellermann said in retrospect, the American government anticipated benefits from such exchanges in three main areas.

First, that exposure to a different environment would produce changes in the views and attitudes of the visitors, enrich their knowledge and skills, and, with it, raise prospects for personal improvement; second, that these changes might cause them to share their (favorable, one hoped) impressions with others, thus contributing to better understanding and improvement of relations between host and home country; and, third, that the German participants would apply the benefits of their experiences by initiating or stimulating actions upon their return, which, in turn, would generate political, social, and educational changes.[9]

The visits were an introduction into the American way of life, conducted with the hope that at least parts of this lifestyle would be transferred to or adopted in Germany, and through this, new allies could be won in Germany. Between 1947 and 1956, more than 12,000 Germans took part in the exchange programs that were first organized and financed by American military and civil authorities including some private institutions as the American Field Service, some foundations and associations, and later in cooperation with German authorities in the formerly occupied German territories. More than 5,500 "leaders" (e.g., educators, politicians, union leaders, journalists, clergy, and women's representatives), 4,100 adolescents (apprentices, students), and over 2,200 university students spent between a few days or weeks and one year in the United States.[10] In the mid-1950s, the Fulbright Program, named after Senator J. William Fulbright, expanded to include Germany. Between 1953 and 1956, this program sent an additional 200 German students to the United States.[11] These "allies" were supposed to form a (transatlantic) elite capable of disseminating their experiences and insights within Germany. The German historian Hermann-Josef Rupieper concluded that the participants of the exchange programs shared their experiences of the American lifestyle with more than three million Germans via personal contacts, public lectures, contacts in schools and universities, or in their function as leaders of public opinion.[12] Two examples will serve to illustrate those numbers.

In 1950, Peter M. Roeder, a teacher at a small village school in Hesse, went to study literature and political science at Sheperd College, Sheperdstown, West Virginia. In an interview in 2004, he remarked that after his immediate return to Germany about "70, 80 people learnt quite a lot from me about America." Besides his studies, he took the opportunity to visit schools in Charleston, Martinsburg, and Washington, DC, where he received a lasting impression of the everyday school life including the style of teaching and the teacher-student relations. "That was really a contrast [to the situation in Germany], which made me think," something that came up again and again in subsequent discussions, Roeder mentioned.[13] In 1966, Roeder became professor of pedagogy at the University of Hamburg and later director at the Max Planck Institute for Human Development in West Berlin, the leading German institute for educational research, and a main advocate of educational and school reform.

Heinrich Roth, a leading German educator and psychologist, went to America on a seven-month trip in 1950 by the invitation of the American Association of Colleges for Teacher Education. During this trip, financed by the Rockefeller Foundation, Roth studied teacher education and child psychology at the University of Maryland, at the Wilson Teachers College, Washington, DC and the State Teachers College at Oneonta, NY. For him, the trip to the United States was an "experience of conversion." In the diary he wrote during the trip, a famous quote from Goethe appeared several times: "*Amerika, Du hast es besser...*" ("America, you're better off..."). Roth was impressed by the "dynamic, lively community" and the fact that in America, "democracy is not only a political system, but a way of life." After his return to Germany, he emphatically stated that in America he had witnessed a pedagogical style that had been characteristic of Germany in the 1920s. "Germany tastes cramped and anxious in comparison."[14] Between 1956 and 1966, Roth was professor of pedagogical psychology in Frankfurt/Main and from 1966 onward, he taught at the University of Göttingen. During this period, Roth became a member of the *Bildungsrat*, a research and advisory institution in the field of education founded by the German federal and state governments.

The experiences of exchange participants such as Roth and Roeder shaped the life and work of many Germans active in the field of education, as teachers in schools, in the education of teachers, and in educational research. The American way of life in education became part of their academic life and influenced their teaching and research in the field. However, despite the fact that thousands of Germans had discovered the democracy in action in America, had widely shared their knowledge and in most cases their appreciation, and even though Germany could claim to have democratic political structures and institutions, many Germans in the 1950s still commented that a "democratic spirit" was lacking in their country.

Exchange Programs and the "American Way of Life" in Education to Be Continued—The "German Educators Missions" 1960–71

At the end of the 1950s, Max Horkheimer, the German consultant of the AJC who was responsible for giving reports on the situation in Germany, and the AJC itself noticed the growth of a new antisemitism and neofascism. This development showed that Germany had neglected to critically deal with its own recent past and that the educational institutions, especially schools, had failed to do this job. Among many Germans, antisemitism and neofascism were seen as byproducts of the strong antidemocratic attitudes. For many, democracy was an institutional framework only, and not a way of life that included personal participation and special norms and values of mutual human and social understanding. Horkheimer and the AJC came to the conclusion that the German school system was in need of democratic reform to fight these tendencies. What was necessary was the introduction of special school subjects like social studies or civic education with a strong emphasis on the history of the Third Reich and a new structure, or at least a new school climate that embodied the democratic ideals of equality and equipped students to face the challenges of life in an open and modern society.

Therefore, a further exchange program mainly geared toward German educators was initiated. It "should not concentrate only on civic education for these educators, but should try to show them the work of a comprehensive high school," Horkheimer and the AJC agreed.[15] Hellmut Becker, who had been involved in the German educators missions project and in 1963 had founded the Institute for Educational Research in Berlin, which later became the Max Planck Institute for Human Development, stated this "challenge" is the heart of what Germans could learn from America: "An experience has to be taken up productively, has to be reshaped in a useful way, and has to inform one's own work."[16]

The trips were organized by the Frankfurt-based Institute for Social Research, its Bureau on Political Education [*Studienbüro zur Soziologie der politischen Bildung*], and the AJC, but were financed by the German government; some German and American foundations like the Ford Foundation, the Volkswagen Foundation, and the Thyssen Foundation, the city of Frankfurt/Main, and the state of Hesse. Almost all of the 127 participants were educators whether teachers, school administrators, professors of education or social and political sciences or history, or school book publishers. Although small in numbers, this was a group of very influential people within the West German educational and school sector. Before and after each of the four- to eight-week trips, the participants met with former or future participants, leading members of the AJC and the American Embassy and German authorities mainly in Frankfurt/ Main to discuss the trips, exchange their ideas, and form an elite network referred to as a cadre of enlightenment (*Kader der Aufklärung*). They were positioned to become an active minority within the German society that would lead the democratization process and that would restructure at least the important elements of the German school system along the lines of the "American model of a democratic society."[17]

As a rule, the trips started in New York City, where the organization of the trip was taken over by a partner organization, the Institute of International Education (IIE). After one or two weeks of introduction to the American educational system and its local, regional, and national administration, which included visits with administrators in Washington, DC and visits to (mainly secondary) schools and university departments in and around New York City, the participants were divided up into pairs to travel to several locations throughout the United States. As one participant said, "From kindergarten to the Center for Advanced Studies in the Behavioral Sciences in Palo Alto, we got to know all levels and forms of efforts in social education."[18] These visits included observations in schools and classrooms, at colleges and universities as well as meetings with leading scientists in education, psychology, sociology, and political science. As the IIE reported to the Ford Foundation in December 1960 about the core part of the trip of the first group, "a well-balanced program was provided…including visits to the University of Colorado at Boulder and Denver University; a lengthy stay at the Curriculum-Making Center at Colorado State College in Creely[sic]; observation of citizenship training in the Denver public schools; observation at the Adult Education Council in Denver; [and] visits to various school systems in Southern Colorado where the majority of students are Spanish-American."[19]

The "American Way of Life" in Education Imagined and Experienced, or a Smile Instead of a Smirk

The American way of life in education was perceived in three different ways. First, the American way of life was seen to generate a social climate conducive to a democracy in action that surrounded and per-meated kindergartens, elementary, and high schools, as well as uni-versities. Second, the American way of life was observed as a specific social climate within the schools/colleges and especially the relation-ship between teachers and students. Third, the American way of life was seen to have unfavorable side effects for education.

Democratic Social Climate

In the fall of 1962, at a conference designed to prepare a group of six educators for their study trip to the United States, Adorno, who had spent many years in exile in the United States, tried to explain a basic American attitude. He emphasized that "adjustment" was an American perspective fundamental to an internal sense of self and self-in-relation as well as to public behavior. Europeans and especially Germans often perceived this attitude of adjustment as superficial and even insincere. Adorno claimed: "The concept of 'adjustment'...must not be seen from its dark side only. Even the conversation of 'keep smiling'—which Europeans will keep smirking at—is better able to lead people to a certain humanity toward others than is that presumptive identity to one's self and that warmth and profundity which much of the time are not there anyway. The experience of America should help anyone to realize that the conditions of his own existence, from the most primi-tive matter to the most sublime, which are taken for granted and have become second nature, are by no means such a matter of course in actual fact."[20]

Adorno described "adjustment" as the ability to imagine oneself in somebody else's shoes, an ability that permits a change of perspective and therefore helps to prevent the exclusivity of the German ideal of "warmth and profundity," which in effect leads to an attitude of igno-rance, superiority, and presumed authority. Max Horkheimer added that for Americans, "respect for the other person's liberty is natural and self-understood." This respect implies frankness, and a "liberal spiri-tual climate" that "manifests itself in the American's ability to laugh at themselves...whereas we [Germans] often lack a genuine sense of humor."[21]

The American "adjustment," "respect," and "liberal spiritual cli-mate" were accompanied by a climate of "education-mindedness" and "community-mindedness" of the American people, as Dietrich Goldschmidt, professor of education at the Teachers College in West Berlin, noted. Goldschmidt recounted his positive impression that the American school was "a social melting pot" and he praised "the desirability of a full school day for the development of democratic attitudes."[22] In a somewhat Tocquevillian manner of traveling around and observing "Democracy in America," the German visitors encountered an open, communicative, and friendly—or in other words: democratic—atmosphere that was characteristic for the American way of life as it informed the everyday life in schools and colleges.

School Climate

The travel reports and the discussions at the meetings before and after the study trips largely focused on evaluating the achievement of the comprehensive school model in U.S. schools. Here, all students were taught in one and the same school, regardless of their achievement level. This was foreign to the German visitors, who were used to the three-tiered German school system where a student's achievement during the first four years of school determined which type of school he or she would progress to following the fifth grade. The observations on this second level mainly concentrated on the comprehensive school structure and the social climate, the communicative patterns, and the self-confidence of students it enabled.

In October of 1961, Friedrich Minssen, a school administrator, who served as the head of the Bureau on Political Education in Frankfurt/Main and who was also one of the organizers of the study trips to the United States, stated that for the political science teacher, "social learning" rather than "social studies" was the "paramount American experience." The transcript of his speech at the preparatory conference for the second study trip read, "[g]enerally speaking, everyday practice of human relationship would prove to be more fruitful than a study of the ways in which knowledge regarding the formal function of democracy is taught. A case in point...was the exemplary relationship between teachers and students."[23] From Minssen's perspective, the social climate in American schools and classrooms was a central focus of the observations and experiences during the study trips. As a rule, he supposed, for most participants, these experiences were the main reason to travel to the United States and to continue to organize the trips. In his internal paper from January of 1966 on the effects of the study trips, he quoted a statement that was representative of the opinions of many participants, "I'm going to use the impetus from the United States especially to dissolve hierarchical and autocratic teaching methods."[24] Hans Graf, a grammar school teacher, who traveled to the United States in the spring of 1964, elaborated on this point. After his return from the study trip, he reported that he had experienced American high schools as "free of hierarchical tendencies between teachers and students and between teachers and the administration. The American teacher is in general more practical than theoretical and tends more towards a *vita activa* than to a *vita contemplativa.*"[25]

Hermann Glaser, who had participated in a study trip in the spring of 1963, described the school visits as "a concrete introduction to a system of schooling, and particularly—which became an important aim for me as well—to a system free of or at least relatively free of compulsion."[26] Glaser, an elementary school teacher, who became the head of the local Department for School and Culture after his return to Nuremberg and was in charge of one of the few Bavarian comprehensive school experiments in the 1970s, argued that in comparison to Germany, America "had been successful in realizing the idea of a comprehensive education for its citizens. Germany does not possess any educational institution comparable to the American high school in the depth and breath of its work in building a nation."[27] Dieter Sauberzweig, the senator for Culture in West Berlin during the late 1970s, had visited American high schools together with Glaser and explained that "greater openness," the "extent of their flexibility," and "respectful conduct" were their main characteristics. He perceived the "school as a living organism" with its participation of students and parents. He mused retrospectively that with libraries and an impressive array of extracurricular activities, it "seemed indeed worth some trouble to transfer some of those aspects into the German school system."[28] Reinhardt Tausch, a professor of education who traveled to the United States in the spring of 1962, reported that in America, he had seen, "for the first time in my life, the social climate in schools" that he and his colleagues had tried to convey through their own teaching to the future teachers they taught.[29] "Constant friendliness and patience" were the main features of the teachers' behavior toward the students.[30]

A final anecdote, which even found its way into *Newsweek*, will demonstrate that it was not the structure of the American school system or its egalitarian image that inspired the greatest learning experience for the German visitors. It was instead the atmosphere in the schools, the relationship between teachers and students, the self-confidence of the students, and their astuteness in all social and civic matters regardless of their economic or racial background, that provided the most profound teaching moments. When *Newsweek* reported about the study trips of German educators to the United States in January of 1965, they began their article with the following passage, "During a recent tour of U.S. schools, Erhard Dornberg, a 41-year-old history teacher from Düsseldorf, Germany, was proudly placed on display by a University of Denver professor. 'Please tell the class how many years of Latin you had in your high school,' said the professor. 'Nine,' said Dornberg. 'How many years of higher mathematics?' 'Eight.' 'Physics and chemistry?' 'Six.' Finally an unimpressed student asked: 'Would you mind telling us, sir, how many years of Hitler you had?' "[31]

This example illustrates what it was that impressed the German teacher and many of his colleagues with the American way of life in education and its results. The American student was totally unimpressed by the German teacher's history of rigorous academic training. Instead, the German visitor was confronted with the courage and wit of a young American who questioned not only his self-understanding but also his concept of education (*Bildung*) and the way it is conveyed in German schools. With his question, the American student not only demonstrated his insight into the relationship between education and social responsibility, but he also showed self-confidence and the ability to think critically. It was understood as the expression and outcome of the democratic patterns he had encountered in school that best prepared students for a democratic way of life.

Unfavorable Side Effects

To have their basic convictions challenged in the way described above was a shattering experience for many Germans. Would letting go of German "profundity" and "seriousness" pave the way for a democratic way of life, would it change the authoritarian, antidemocratic climate in German schools? Furthermore, would it allow for the formation of young, democratic minds that would be the foundation of a democratic society? Already in 1962, the reports about the American "school of happiness and easiness,"[32] as a leading German newspaper wrote in July of that year, were countered by some other observations that were also recorded in the travel reports and shadowed by American foreign and domestic policies in the 1960s. Ever since the program had started in the fall of 1960, the German visitors noted and remarked upon some of the unfavorable effects of the American way of life in education. Diedrich Goldschmidt observed that many schools he visited "were either too large or too small." The "inherent educational aim of encouraging students to challenge society" could not always be realized and the "overabundance of extra-curricular activities" merely created an "additional field for competition." He added that in some cases the quality of classroom teaching was poor. In particular, he observed too much of an emphasis on "true and false" and in part a "weak handling of controversial issues" in the classroom.[33] In July of 1962, Gernot Koneffke, lecturer and later professor of education at teachers colleges in Hesse and Lower Saxony, pointed out that there seemed to be a central problem with "certain articles of faith, the validity of which is taken to be above dispute." This "prevailing pragmatism," Koneffke noted, "seems to deprive Americans of the possibility of criticizing their social structures."[34]

Kurt Sontheimer's introductory statement at the summary conference in New York City in May of 1964 was even more poignant, "[T]he 'American way of life' is, in reality, a myth." The professor of political science at the Free University of Berlin continued, "[t]he 'American Dream' of equality is conveyed in classrooms, but reality, for example the truth about class structure and racial conflict, is not taught. This oversight of 'real' problems may serve as the means of keeping society 'together,' but should 'real' problems continue to be ignored? In the United States, the teaching of social studies on the secondary level is not as good as it is in Germany."[35] The ensuing dialogue between Max Birnbaum, educational consultant of the AJC, and Kurt Sontheimer highlighted the difference between the West German and the American approaches to education. Birnbaum first defended the American system by referring to the "wide range of quality of teaching in this field." He then went on to emphasize that the final "result of this teaching, and of any education, is not so much what is learned in a classroom but how the student behaves as a person and what he contributes to society upon the completion of his formal education."[36] To this, Sontheimer replied that "German and American educators differ in terms of the purpose of education." Germans did not think of schools as "social agent[s], but rather as a means to convey the tools and skills which will enable a student to exist in society." German teachers were "not concerned with the behavioral consequences of the educative process on the student." Alluding to the sputnik shock, Sontheimer finally stated, "[p]erhaps there is not enough emphasis in American schools on the acquisition of knowledge and the attainment of intellectual competence."[37] In his reply, Herbert Schueler, director of the department of education at Hunter College in New York City, made evident the miserable side of the American education at this time. American education and the American school system were in a "period of transition." But now and in the future, "it seeks to obtain a balance between the mastering of skills and the acquisition of knowledge and the retention of concern of the total development of the individual." Schueler finished his statement on the means and ends of American education by referring to the social function of schooling in the best tradition of progressive education: "The social aspects of life cannot be separated from school life. If the schools are not social agencies, it is time that they become life-centered, at least. Education should be more than 'just learning for the sake of learning.'"[38]

It was not only the Cold War that forced the American system of education into a process of transformation. The class divide within the school system and the unequal treatment of African Americans across society were challenges for the American school system as well. The German visitors, who traveled through the country during the 1960s, took notice of these developments as well.[39] Hartmut von Hentig, professor of education at the University of Göttingen, had studied at Elisabeth College, Elisabethtown, PA, and at the University of Chicago[40] between 1948 and 1953 and went on a study trip to the United States with the German educators mission in the spring of 1967. In retrospect, he noted that the Vietnam War had made many Germans skeptical about the American way of life, and that no one really dealt with the complexity of this problem in relation to education.[41]

The American side slowly started to pull out of the exchange program and their engagement in reeducating Germany. Although in April of 1966, an internal paper of the Ford Foundation, which co-financed the program, assessed the German educational and school system as not "safe" and stated that "America has failed to bring German education along,"[42] the Ford Foundation did not renew their grant. The city of Frankfurt/Main and Hesse followed suit, and in the end, the funds of the German federal government were cut as well. The program ended in 1971. At the end of 1966, the AJC already noted that "foreign influence is losing more and more of its effectiveness on the German scene," with one AJC program officer reporting that he felt that "German leaders [were] now more ready for action than before and for the good reason that they themselves are afraid of the future."[43]

The main aim of the exchange program had been the education and strengthening of influential people within the German educational and school sector, so that they in turn would democratize schools in Germany. Within the discourse on educational and school reform, the contrast between the supposedly undemocratic character of German schools and the democratic spirit within American schools was a strong argument for change in Germany. It was not so much the structure of the American school system that was seen as the heart of the American way of life in education. It was the inner structure, the spirit of cooperation in teacher-student-relations, the nonauthoritarian style of teaching, the pluralism in organizing and leading the educational process to create a "democratic climate" of teaching and learning, as well as the social function that schools served for society at large—that had impressed the German visitors.

Conclusion

Even though the "foreign influence" on German education and schooling was receding in the 1960s, the "foreign example" and especially the American way of life in education continued to shape the German educational reform process and reform discourse, as well as German educational practices.

Many Germans, who had participated in the exchange programs after World War II, later became members of the German *Bildungsrat*, the most central educational research and advisory institution founded by the German federal and state governments between 1965 and 1975. Among those serving on the *Bildungsrat* were Heinrich Roth, Peter M. Roeder, Hartmut von Hentig, Dietrich Goldschmidt, as well as Wolfgang Edelstein, one of the directors of the Max Planck Institute for Human Development, Jakob Muth, professor of education North Rhine-Westphalia, Ludwig von Friedeburg, Minister of Education and Culture in Hesse, and Hildegard Hamm-Brücher, permanent secretary at the Federal Ministry of Education. Hellmut Becker, Hans Schütte, Minister of Education and Culture in Hesse in the mid-1960s, and Carl-Heinz Evers, Senator of Education in West Berlin, who were involved in selecting "suitable" candidates (*Kader der Aufklärung*) for the German educators missions, were also members of the *Bildungsrat*. Some of them published widely on the American educational and school system and reform (e.g., Hentig, Edelstein, Roeder) and assimilated American research in the field (e.g., Roth, Roeder, Hentig, Goldschmidt).[44]

In the early 1960s, the "foreign example" of reform and development in the educational and school sector played a crucial role in the discourse on school reform in Germany. At the same time as the Organisation for Economic Co-operation and Development (OECD) was conducting its first comparative international survey in the field of education, the German Conference of the State Ministers of Education and Culture (KMK), the central institution to coordinate the different German state policies in education, was also referring to the reform plans in certain OECD countries like England, Sweden, and some others.[45] America was held up as an example primarily in the field of research, but also with regard to school structure, democratic school climate, and experimental pragmatism in education and schooling.[46] The "international example" and the positive image of the American way of life in education thus entered the German reform discourse as "additional meaning" and provided a helpful starting point for a scientific and international reevaluation of genuinely German positions and arguments.[47]

The American way of life not only influenced the theoretical discourse on school reform, but it also informed the way theoretical insights were put into practice. For example, the German model of the comprehensive school, as it was introduced in some German states from the mid-1960s onward, following the recommendation of the *Bildungsrat*, clearly showed the American influence in its structure, organization, and in its aim to create a democratic school climate. The same influence was visible in some other reform school projects that were started in the early 1970s like the *Laborschule* and the *Oberstufen-Kolleg* in Bielefeld lead by Hartmut von Hentig.[48] Hentig had participated in the German educators missions in 1967 because he had "wanted to start his own school" upon his return to Germany.[49] Other reform elements like the "course system," "student councils," "team teaching," "social learning," or "civic education" have become commonplace in German schools, teacher education and educational research, but when they were first introduced in the late 1960s and early 1970s, their American roots were more or less undeniable. However, the American way of life was not by far the only international influence on the German school reform discourse. In public perception, other international models like the Swedish or English model of comprehensive schooling became more important than the American example.[50] In addition, the negative image that the American way of life had acquired by the late 1960s at the latest tainted the initially positive reception of American innovations in educational and school reform. Both of those developments concurred in eclipsing the influence of the American way of life on the German school reform process. So, even though many elements of the German school reform had actually been inspired by the American example, their actual origin was purposefully concealed or simply "hidden."

Notes

1. Report on the Summary Session German Educators Program, May 23, 1963, appendix, p. 3, Horkheimer's bequest, V 190, Archivzentrum Frankfurt/Main.

2. See Charles Dorn and Brian Puaca (this volume); also, inter alia, Ute Gerhardt, "Re-Education als Demokratisierung der Gesellschaft Deutschlands durch das amerikanische Besetzungsregime. Ein historischer Bericht," *Leviathan* 27 (1999): 355–385; Ute Gerhard, *Soziologie der Stunde Null*. *Zur Gesellschaftskonzeption des amerikanischen Besatzungsregimes in Deutschland 1944–1945/46* (Frankfurt/Main: Suhrkamp, 2005), 101–105, 260–264; James F. Tent, *Mission on the Rhine*. *Reeducation and Denazification in American-Occupied Germany* (Chicago/London: University of Chicago Press, 1982); Karl-Ernst Bungenstab, *Umerziehung zur Demokratie? Re-education-Politik im Bildungswesen der US-Zone 1945–1949* (Düsseldorf: Bertelsmann Verlag, 1970); Karl-Heinz Füssl, *Die Umerziehung der Deutschen. Jugend und Schule unter den Siegermächten des Zweiten Weltkriegs 1945–1955* (Paderborn et al.: Schöningh, 1994); Beate Rosenzweig, *Erziehung zur Demokratie? Amerikanische Besatzungs- und Schulreformpolitik in Deutschland und Japan* (Stuttgart: Klett, 1998).

3. John Slawson, *New York Post*, September 17, 1963. In the newspaper article, Slawson referred to his visit to West Germany in 1959.

4. Thomas Koinzer, "Das pädagogische Amerika in der Weimarer Republik. Rezeption und Externalisation der 'Schule der Demokratie,'" in *Mythos USA. "Amerikanisierung" in Deutschland seit 1900*, ed. Frank Becker and Elke Reinhardt-Becker (Frankfurt/Main./New York: Campus Verlag, 2006), 135–150.

5. David Tyack, *Seeking Common Ground. Public Schools in a Diverse Society* (Cambridge, MA: Harvard University Press, 2003), 32.

6. See, inter alia, Klaus Hüfner et al., *Hochkonjunktur und Flaute: Bildungspolitik in der Bundesrepublik Deutschland 1967–1980* (Stuttgart: Ernst Klett Verlag, 1986), 149–212. Heinz-Elmar Tenorth, *Geschichte der Erziehung: Einführung in die Grundzüge ihrer neuzeitlichen Entwicklung* (Weinheim: Juventa Verlag, 2000), 293–313.

7. "*Zusatzsinn*" see Jürgen Schriewer, "Problemdimensionen sozialwissenschaftlicher Komparatistik," in *Vergleich und Transfer. Komparatistik in den Sozial-, Geschichts- und Kulturwissenschaften*, ed. Hartmut Kaelble and Jürgen Schriewer (Frankfurt/Main: Campus Verlag, 2003), 49.

8. Henry J. Kellermann, *Cultural Relations as an Instrument of U.S. Foreign Policy. The Educational Exchange Program between the United States and Germany 1945–1954* (Washington, DC: U.S. Department of State, 1978), 211.

9. Ibid.

10. Ellen Latzin, *Lernen von Amerika? Das U.S.-Kulturaustauschprogramm für Bayern und seine Absolventen* (Stuttgart: Franz Steiner Verlag, 2005), 83–84; inter alia, Tent, *Mission on the Rhine*, 111–176; Annette Puckhaber, "German Student Exchange Programs in the United States, 1946–1952," *Bulletin of the German Historical Institute Washington* 30 (2002): 123–141; Brian Puaca, *Missionaries of Goodwill: German Exchange Teachers and American Democratic Society after World War II* (Typoskript, 2003); Karl-Heinz Füssl, *Deutsch-amerikanischer Kulturaustausch im 20. Jahrhundert* (Frankfurt/Main: Campus Verlag, 2004), 186–201; Oliver Schmidt, "Small Atlantic World: U.S. Philanthropy and the Expanding International Exchange of Scholars after 1945," in *Culture and International History*, Jessica C.E. Gienow-Hecht and Frank Schumacher, eds. (New York/Oxford: Berghahn, 2003), 115–134.

11. Ulrich Littmann, *Gute Partner—Schwierige Partner. Anmerkungen zur akademischen Mobilität zwischen Deutschland und den Vereinigten Staaten von Amerika 1923–1993* (Bonn: DAAD-Forum, 1996), 83–86. Latzin, *Lernen von Amerika?* 432. Wolfgang C. Müller, *Wie Helfen zum Beruf wurde. Band 2. Eine Methodengeschichte der Sozialarbeit 1945–1995* (Weinheim: Beltz Verlag, 1997), 42–51. Füssl, *Deutsch-amerikanischer Kulturaustausch im 20. Jahrhundert*, 237–270.

12. Hermann-Josef Rupieper, *Die Wurzeln der deutschen Nachkriegsdemokratie. Der amerikanische Beitrag 1945–1952* (Opladen: Westdeutscher Verlag, 1993), 399–400.
13. Interview with the author, February 2004.
14. Cit. Bibliothek für Bildungshistorische Forschung des Deutschen Instituts für Internationale Pädagogische Forschung (BBF), ed., *Realistisches Denken verlangt geisteswissenschaftlichen Kontext. Prof. Dr. Heinrich Roth zum 100. Geburtstag* (Berlin: BBF, 2006), 26–27.
15. Letter Slawson to Horkheimer, July 25, 1961, FAD-1, Foreign Countries, Record Group 347.7.1, Germany, Box 40, AJC archive.
16. Report on the conference, January 14, 1961, p. 12, Horkheimer's bequest, IX 235, 1–50, Archivzentrum Frankfurt/Main.
17. Clemens Albrecht et al., *Die intellektuelle Gründung der Bundesrepublik. Eine Wirkungsgeschichte der Frankfurter Schule* (Frankfurt/Main: Campus Verlag, 1999), 443.
18. Report on the conference, January 14, 1961, p. 2, Horkheimer's bequest, IX 235, 1–50, Archivzentrum Frankfurt/Main.
19. Final Report on the Ford Foundation, October 1960, p. 2 Horkheimer's bequest, IX 235, 1–50, 12, Archivzentrum Frankfurt/Main.
20. Adorno at the conference in Frankfurt/Main, October 11, 1962, Horkheimer's bequest, IX 235, 51–71, 61, Archivzentrum Frankfurt/Main.
21. Summary report, October 16, 1961, p. 1, Horkheimer's bequest, IX 235, 1–50, 29, Archivzentrum Frankfurt/Main.
22. Final Report on the Ford Foundation, October 1960, p. 2, Horkheimer's bequest, IX 235, 1–50, 12, Archivzentrum Frankfurt/Main.
23. Summary report, October 16 1961, p. 2, Horkheimer's bequest, IX 235, 1–50, 29, Archivzentrum Frankfurt/Main.
24. Typescript by Friedrich Minssen, January 1966, p. 5, Horkheimer's bequest, V 187, 207, Archivzentrum Frankfurt/Main.
25. Protocol of the conference in Frankfurt/Main, October 8, 1964, p. 3, Horkheimer's bequest, V 190, Archivzentrum Frankfurt/Main.
26. Interview with the author, May 2004.
27. Protocol of the conference in Frankfurt/Main, July 6, 1963, p. 2, Horkheimer's bequest, V 190, Archivzentrum Frankfurt/Main.
28. Interview with the author, June 2004.
29. Typescript by Friedrich Minssen, January 1966, p. 2, Horkheimer's bequest, V 187, 207, Archivzentrum Frankfurt/Main.
30. Protocol of the Evaluation Conference, July 3, 1962, p. 4, FAD-1, Foreign Countries, Record Group 347.7.1, Germany, YIVO Institute for Jewish Research, Archive.
31. *Newsweek*, January 18, 1965, p. 53.
32. Frankfurter Allgemeine Zeitung July 30, 1962.
33. Final Report on the Ford Foundation, October 1960, p. 2, Horkheimer's bequest, IX 235, 1–50, 12, Archivzentrum Frankfurt/Main.
34. Protocol of the Evaluation Conference, July 3, 1962, p. 8, FAD-1, Foreign Countries, Record Group 347.7.1, Germany, YIVO Institute for Jewish Research, Archive.
35. Report on summary session, May 23, 1963, p. 1, Horkheimer's bequest, V 190, Archivzentrum Frankfurt/Main.
36. Ibid., 4.
37. Ibid., 5.
38. Ibid., 5. James Bryant Conant, *The American High School Today: A First Report to Interested Citizens* (New York: McGraw-Hill, 1959) (German edition in 1960, trans. and ed. by Hartmut von Hentig); Steve Mintz, *Huck's Raft. A History of American Childhood* (Cambridge, MA: Belknap Harvard University Press, 2006), 287–291.
39. Travel report, inter alia, by Wolfgang Bobke, Ministry of Education and Arts in Hesse and editor of the Journal "Gesellschaft, Staat, Erziehung," January 1967, Horkheimer's bequest, V 189, Archivzentrum Frankfurt/Main.

40. Hartmut von Hentig, *Mein Leben—bedacht und bejaht. Kindheit und Jugend* (München: Carl Hanser Verlag, 2007), 287–349. Füssl, *Deutsch-amerikanischer Kulturaustausch im 20. Jahrhundert*, 263–267.
41. Interview with the author, July 2004.
42. Comments on the Program of the Study Bureau on Political Education, Frankfurt Institute of Social Research, Memorandum from J. R. Huntley to S. T. Gordon, April 18, 1966, RE 2582, PA 63–511, Sec. 4, Ford Foundation Archive.
43. John Slawson to Jacob Blaustein, November 30, 1966, JSX, AJC archive.
44. Thomas Koinzer, *Auf der Suche nach der "demokratischen Schule." Amerikafahrer, Bildungstransfer und Schulreform in der bundesdeutschen Bildungsreformära* (forthcoming).
45. Diedrich Benner and Herwart Kemper, *Theorie und Geschichte der Reformpädagogik. Teil 3.2.: Staatliche Schulreform und reformpädagogische Schulversuche in den westlichen Besatzungszonen und in der BRD* (Weinheim: Beltz Verlag, 2007), 186–187.
46. Ibid., 184–185, 368.
47. Jürgen Schriewer, "Problemdimensionen sozialwissenschaftlicher Komparatistik," in *Vergleich und Transfer. Komparatistik in den Sozial-, Geschichts- und Kulturwissenschaften,* Hartmut Kaelble and Jürgen Schriewer, eds. (Frankfurt/Main: Campus Verlag, 2003), 49; Gita Steiner-Khamsi, "Vergleich und Subtraktion. Das Residuum im Spannungsfeld zwischen Globalem und Lokalem," in *Vergleich und Transfer. Komparatistik in den Sozial-, Geschichts- und Kulturwissenschaften,* Hartmut Kaelble and Jürgen Schriewer, eds. (Frankfurt/Main: Campus Verlag), 377–379.
48. Benner and Kemper, *Theorie und Geschichte der Reformpädagogik. Teil 3.2*, 322–340.
49. Interview with the author, July 2004.
50. Theodor Sander, Hans-G. Rolff and Gertrud Winkler, *Die demokratische Leistungsschule: Zur Begründung und Beschreibung der differenzierten Gesamtschule* (Berlin et al.: Schroedel Verlag, 1967), 86–97.

Bibliography

Albrecht, Clemens et al. *Die intellektuelle Gründung der Bundesrepublik. Eine Wirkungsgeschichte der Frankfurter Schule.* Frankfurt/Main: Campus Verlag, 1999.

Benner, Diedrich and Herwart Kemper. *Theorie und Geschichte der Reformpädagogik. Teil 3.2.: Staatliche Schulreform und reformpädagogische Schulversuche in den westlichen Besatzungszonen und in der BRD.* Weinheim: Beltz Verlag, 2007.

Bibliothek für Bildungshistorische Forschung des Deutschen Instituts für Internationale Pädagogische Forschung (BBF), ed. *Realistisches Denken verlangt geisteswissenschaftlichen Kontext. Prof. Dr. Heinrich Roth zum 100. Geburtstag.* Berlin: BBF, 2006.

Bungenstab, Karl-Ernst. *Umerziehung zur Demokratie? Re-education-Politik im Bildungswesen der US-Zone 1945–1949.* Düsseldorf: Bertelsmann Verlag, 1970.

Conant, James Bryant. *The American High School Today: A First Report to Interested Citizens.* New York: McGraw-Hill, 1959.

Füssl, Karl-Heinz. *Die Umerziehung der Deutschen. Jugend und Schule unter den Siegermächten des Zweiten Weltkriegs 1945–1955.* Paderborn: Schöningh, 1994.

———. *Deutsch-amerikanischer Kulturaustausch im 20. Jahrhundert: Bildung, Wissenschaft, Politik.* Frankfurt/Main: Campus Verlag, 2004.

Gerhardt, Ute. "Re-Education als Demokratisierung der Gesellschaft Deutschlands durch das amerikanische Besetzungsregime. Ein historischer Bericht." *Leviathan* 27 (1999): 355–385.

———. *Soziologie der Stunde Null. Zur Gesellschaftskonzeption des Amerikanischen Besatzungsregimes in Deutschland 1944–1945/46.* Frankfurt/Main: Suhrkamp, 2005.

Hentig, Hartmut von. *Mein Leben—Bedacht und Bejaht. Kindheit und Jugend.* München: Carl Hanser Verlag, 2007.

Hüfner, Klaus. *Hochkonjunktur und Flaute: Bildungspolitik in der Bundesrepublik Deutschland 1967–1980.* Stuttgart: Ernst Klett Verlag, 1986.

Kellermann, Henry J. *Cultural Relations as an Instrument of U.S. Foreign Policy. The Educational Exchange Program between the United States and Germany 1945–1954.* Washington, DC: U.S. Department of State, 1978.

Koinzer, Thomas. "Das pädagogische Amerika in der Weimarer Republik. Rezeption und Externalisation der 'Schule der Demokratie,'" in *Mythos USA: "Amerikanisierung" in Deutschland seit 1900,* ed. Frank Becker and Elke Reinhardt-Becker, 135–150. Frankfurt/ Main: Campus Verlag, 2006.

———. Auf der Suche nach der "demokratischen Schule." Amerikafahrer, Bildungstransfer und Schulreform in der bundesdeutschen Bildungsreformära (forthcoming).

Latzin, Ellen. *Lernen von Amerika? Das U.S.-Kulturaustauschprogramm für Bayern und seine Absolventen.* Stuttgart: Franz Steiner Verlag, 2005.

Littmann, Ulrich. *Gute Partner—Schwierige Partner. Anmerkungen zur akademischen Mobilität zwischen Deutschland und den Vereinigten Staaten von Amerika. 1923–1993.* Bonn: DAAD-Forum, 1996.

Mintz, Steve. *Huck's Raft. A History of American Childhood.* Cambridge, MA: Belknap Harvard University Press, 2006.

Müller, Wolfgang C. *Wie Helfen zum Beruf wurde. Band 2. Eine Methodengeschichte der Sozialarbeit 1945–1995.* Weinheim: Beltz Verlag, 1997.

Puaca, Brian. *Missionaries of Goodwill: German Exchange Teachers and American Democratic Society after World War II.* Manuscript, 2003.

Puckhaber, Annette. "German Student Exchange Programs in the United States, 1946–1952." *Bulletin of the German Historical Institute* 30 (2002): 123–141.

Rosenzweig, Beate. *Erziehung zur Demokratie? Amerikanische Besatzungs- und Schulreformpolitik in Deutschland und Japan.* Stuttgart: Klett, 1998.

Rupieper, Hermann-Josef. *Die Wurzeln der deutschen Nachkriegsdemokratie. Der amerikanische Beitrag 1945–1952.* Opladen: Westdeutscher Verlag, 1993.

Sander, Theodor, Hans-G. Rolff and Gertrud Winkler. *Die demokratische Leistungsschule: Zur Begründung und Beschreibung der differenzierten Gesamtschule.* Berlin: Schroedel Verlag, 1967.

Schmidt, Oliver. "Small Atlantic World: U.S. Philanthropy and the Expanding International Exchange of Scholars after 1945," in *Culture and International History,* ed. Jessica C.E. Gienow-Hecht and Frank Schumacher, 115–134. New York: Berghahn, 2003.

Schriewer, Jürgen. "Problemdimensionen sozialwissenschaftlicher Komparatistik," in *Vergleich und Transfer. Komparatistik in den Sozial-, Geschichts- und Kulturwissenschaften,* ed. Hartmut Kaelble and Jürgen Schriewer, 9–54. Frankfurt/Main: Campus Verlag, 2003.

Steiner-Khamsi, Gita. "Vergleich und Subtraktion. Das Residuum im Spannungsfeld zwischen Globalem und Lokalem," in *Vergleich und Transfer. Komparatistik in den Sozial-, Geschichts- und Kulturwissenschaften,* ed. Hartmut Kaelble and Jürgen Schriewer, 377–379. Frankfurt/Main: Campus Verlag, 2003.

Tenorth, Heinz-Elmar. *Geschichte der Erziehung: Einführung in die Grundzüge ihrer neuzeitlichen Entwicklung.* Weinheim: Juventa Verlag, 2000.

Tent, James F. *Mission on the Rhine. Reeducation and Denazification in American-Occupied Germany.* Chicago, IL: University of Chicago Press, 1982.

Tyack, David. *Seeking Common Ground. Public Schools in a Diverse Society.* Cambridge, MA: Harvard University Press, 2003.

PART IV

After the Cold War, in the Face of Terror

CHAPTER EIGHT

American Academics and Education for Democracy in Post-Communist Europe

LAURA B. PERRY

This chapter presents a critical analysis of American educational reconstruction efforts in post-communist Europe after the end of the Cold War. In the 1990s the Third Wave of worldwide democratization was cresting, and in the field of education, democracy, and democratization enjoyed renewed interest. For many American scholars and researchers, the fall of the Berlin Wall opened up significant opportunities to offer workshops, attend conferences in the region, serve as expert advisors, participate in exchange programs, and compete for substantial grants from the U.S. government to carry out this work. The primary fields of interaction between educators from the East and West were in civic education, pre- and in-service teacher training, and curriculum development. This study will examine these reconstruction efforts by American education scholars with a critical lens that makes use of intellectual history on the social construction of Eastern Europe.

The analysis examines the aims and assumptions of the American scholars who participated in education projects in the region through their academic writing. The data set contains 96 academic journal articles, books, and book chapters written by American scholars about post-communist European schooling between the years 1989 and 2001. Descriptive statistics and qualitative content analysis are used to analyze the documents. The first finding is that American perceptions of post-communist schooling were largely negative and normative, and based on a one-sided transmission of expertise. The second finding is that most of the negative perceptions centered on teaching and learning. The third finding is that reconstruction efforts were justified by the perceived need for educational democratization. The notions of democracy used, however, were limited in such a way as to give superiority to American conceptions. The chapter concludes with the argument that American educational reconstruction efforts in post-communist Europe were a mechanism for constructing an inferior, eastern "other" that served the interests of the American academic and educational community.

Background

Forty years of communism in Central and Eastern Europe, and almost 90 years in the Soviet Union, led to the isolation of education in the region from developments occurring elsewhere in the world. This isolation was seen by educators from both within and outside the region as unfortunate, and soon after the fall of the Berlin Wall in 1989, efforts to dismantle the educational Iron Curtain were initiated. Researchers and academics, policymakers and ministry officials, and administrators and practitioners in the post-communist countries welcomed the opportunity to learn about and from educators in OECD countries, especially Europe and North America.

While educators in the post-communist countries were curious about trends abroad, they did not seek a radical transformation of their educational systems. Rather, experts and the lay public alike generally considered their national systems of education to be sound and at times excellent.[1] These generally favorable opinions derived in part from measures of successful educational outcomes, such as universal literacy, low dropout rates from secondary education and strong achievement on international tests of student achievement such as the Trends in International Mathematics and Science Survey (TIMSS).[2]

After communist rule ended in 1989–90, most countries in the region initiated a series of significant educational reforms. These reforms centered primarily on increasing diversity and choice into systems that had been highly centralized and standardized. The main reforms were: (1) depoliticizing education (i.e., removing the mandatory teaching of Marxism-Leninism; (2) removing the state's monopoly on education by allowing private or nongovernment schools to be established; (3) increasing parents' and students' ability to choose a school; and (4) decentralizing educational governance by giving more autonomy to local and regional authorities.[3]

These four main reforms radically changed the structure and governance of education in the post-communist countries. They did not, however, lead to significant change within schools and classrooms.[4] Many teachers have adopted more pupil-centered instructional strategies, and curriculum in a few subjects, such as civic education and history, has been overhauled. Overall, however, teaching and learning have not changed dramatically since the end of communism. Nor has the education profession, government bodies, or the lay public demanded profound changes to teaching and learning.

This does not mean, however, that educators in the post-communist countries are not interested in adopting new instructional techniques or learning about external trends and practices, or that they are resistant to change.[5] Most teachers and educators do not advocate a wholesale break with the past or exhibit a dogged resistance to change. Rather, they are proud of their accomplishments and contributions, and at the same time seek ways to improve their practice.[6]

Decades of isolation and professional curiosity led many educators and researchers to welcome collaborative projects, exchange visits, workshops, and study opportunities, and professional development projects with colleagues from Western Europe and North America. Hundreds of academics and researchers, as well as teachers and principals, from Western Europe and North America have led projects in post-communist countries. For example, from 1998 to 2006 (the years for which data is publicly available), the U.S. government sent over 60 Fulbright scholars in the field of education to teach or conduct research in a post-communist European country.[7] Most of the 96 documents included in this study were written by American academics who had visited and worked, if only briefly, in the region, and only a few of these authors were Fulbright scholars.

Policy makers were also engaged in extension collaboration with foreign partners, especially with European organizations. Significant sources of foreign influence came mostly from the European Union, the Organization for Economic Co-operation and Development (OECD), and the Council of Europe (COE), all three of which are multilateral organizations based in Europe.[8] While the United States is a member country of the OECD, the organization's headquarters are based in Paris and their education projects and publications are led by European researchers and policy makers. American bilateral or American-based multilateral organizations, such as the World Bank, played a much smaller role in educational reconstruction efforts in the post-communist countries.

By contrast, most American actors involved in reconstruction efforts in the post-communist countries were individual academics, researchers, and practitioners. Sometimes they received grants from the U.S. government to fund their projects. As mentioned earlier, a small number of these individuals received funding from the government's Fulbright scholar program to teach in the region. Many of the American academics involved in projects in the post-communist countries published work about their experiences in journals or books, and it is on this body of work that the chapter's analysis is based.

Probably the largest American-based organization directly involved in educational reconstruction efforts in the post-communist region is George Soros' Open Society Institute (OSI), a nongovernmental organization based in New York. The OSI funds projects in many countries, including the former communist countries of Eastern and Central Europe. Their priorities in the region are higher education and education for vulnerable and marginalized children, especially Roma. While the OSI has been actively involved in education reconstruction and reform, an analysis of their projects and their impacts would deserve a separate treatment, and has therefore been excluded from this study. Another reason to exclude the OSI from this chapter is that George Soros is a Hungarian-born immigrant who has remained committed to his home country and the larger region. As such, the OSI can not be considered a purely American agent of educational reconstruction.

Context

In this section I briefly describe the social, political, and educational context of the 13 countries included in the study. I also compare the post-communist countries with Germany, its biggest and nearest neighbor to the West, and the United States since the study concerns American scholars' reconstruction efforts in the region.

Together with the other countries of Europe, most of the 13 countries belong to multilateral organizations that require a commitment to democratic governance, human rights, and market economy principles, such as the COE, North Atlantic Treaty Organization (NATO), OECD, and the EU. During 1989–2001, the time period of the dataset, all 13 countries had joined the COE; 7 joined in 1993, with the remaining 3 joining by 1996. The COE's primary aim is to "protect human rights, pluralist democracy and the rule of law" and is the oldest multilateral political organization in Europe.[9] Also during the time period of the data, 3 of the countries (the Czech Republic, Poland, and Hungary) had joined NATO (another 7 of the 13 countries joined in 2004). By 1996, 3 of the 13 countries had joined the OECD; a fourth joined in 2000. Eight of the 13 countries were being "fast-tracked" for full membership in the European Union (8 joined in 2004, and another 2 joined in 2007). Of the 13 countries, the Czech Republic, Poland, and Hungary have been at the forefront of joining these multilateral organizations. These three countries made the fastest strides toward implementing a political democracy and market economy after 1989. At the other end of the spectrum, Russia, Belarus, and Ukraine have experienced many more challenges and are still struggling to become full democracies. Of the four organizations mentioned here, they are members only of the COE. Somewhere in the middle are countries such as Romania and Bulgaria.

Politically, most of the post-communist countries enjoy the same democratic freedoms as more established democracies, including Germany and the United States. Freedom House, a nonprofit and nonpartisan organization that monitors democracy around world, annually evaluates countries based on their political and civil liberties, then average the scores into three broad categories: free, partly free, and not free.[10] The Czech Republic, Hungary, Poland, Estonia, Lithuania, Latvia, and Slovenia became "free" immediately after the end of communism or Soviet rule; by 1996, Bulgaria, Estonia, Slovakia, and Romania had become free as well. Ukraine did not receive a "free" ranking until 2005, and Belarus and Russia have actually become less free since the fall of communism. Thus, by the early 1990s, most of the countries in the sample were just as democratic as their neighbors to the west, and by the mid-1990s all but three were fully democratic.

The post-communist countries were (and continue to be) comparable with the United States and Germany on educational indicators during the timeframe of the study. All 13 countries have literacy rates at 99 percent or above, as do almost all OECD countries.[11] Students in the post-communist countries perform similarly with their peers in Germany and the United States on international achievement tests. Compared to students in the United States, only two post-communist countries (Lithuania and Romania) had a lower performance on the eighth grade math component of TIMSS in 1999; four post-communist countries (Slovakia, Hungary, Slovenia, and Russia) scored significantly higher, and another three (Czech Republic, Bulgaria and Latvia) showed no statistical difference (Germany did not participate in TIMSS 1999).[12] On the science component of TIMSS 1999, the same two countries scored lower than the United States, with another four scoring higher (Slovakia, Hungary, Slovenia, and the Czech Republic); three countries (Russia, Bulgaria, and Latvia) showed no statistical difference.

The OECD's Programme in International Student Assessment (PISA) tests the ability of 15-year-olds to apply knowledge and solve problems. Five of the 13 countries included in this study participated in PISA 2000, the first year that it was conducted. Russia and Latvia both performed lower than Germany or the United States, Poland, and Hungary performed at the same level as the two comparison countries, and the Czech Republic scored higher than Germany (but not statistically differently than the United States).

In summary, on these broad indicators the post-communist countries are comparable to their peers to the west. Certainly their economies continue to face more challenges, and citizens have less buying power and discretionary funds. Beyond this, however, it is hard to argue that the post-communist countries were or are significantly less developed than OECD countries. The three countries exhibiting the largest differences are Ukraine, Belarus, and Russia, all of which have faced serious obstacles in transitioning from communism to a transparent market economy and democratic governance.

Method

The chapter analyses documents written by American academics involved in educational reconstruction efforts in post-communist Europe and which were published between 1989 and 2001. Documents were limited to academic sources including journal articles, books, and book chapters. Articles in newspapers serving the academic community, such as *The Chronicle of Higher Education*, were not included in the analysis since the authors are typically staff journalists reporting on projects rather than academics involved in educational reconstruction. Articles that had a non-American coauthor were also omitted.

Analyzing academic writings as a primary source of data is common in the disciplines of history, literature, and post-colonial studies. In his groundbreaking book *Orientalism*, Edward Said analyzed Western scholarship and other written works (reports, diaries, letters, etc.) by intellectuals and academics to explore the relationship between knowledge and power and the West's imperialism in the orient (Middle East).[13] This study uses Said's method as well as his theoretical insights as a conceptual framework.

Other than four Balkan countries (Albania, Croatia, Bosnia-Herzogovina, Serbia), all of the post-communist countries in Europe were included in the data search. These four countries were omitted since education during the time period of the analysis was severely disrupted due to war. Instead, analysis of American reconstruction efforts in the Balkans would need a separate treatment, as it has indeed been provided in this volume. Slovenia, however, avoided involvement in the war, and therefore it is included in the analysis. Also omitted from the study were the post-communist countries of central Asia, such as Kazakhstan. The central Asian countries are incommensurable because of their relatively recent history as nation-states, and their markedly different culture and level of economic development. In fact, the only thing the newly independent states of central Asia share with the other post-communist countries is a history of Soviet rule.

After outlier countries were removed, 13 post-communist European countries remained: Poland, Czech Republic, Slovakia, Hungary, Slovenia, Bulgaria, Romania, Lithuania, Latvia, Estonia, Ukraine, Belarus, and Russia. The documents were found through the standard bibliographic databases and search engines, as well as cross checking reference lists and bibliographies. While I cannot guarantee that the data set is 100 percent complete, that is, that it contains every single piece written by an American scholar about post-communist European education, it is comprehensive and most likely is near the saturation point.

The study primarily uses qualitative content analysis to interpret written texts, along with descriptive statistics to give weight to the findings. For example, I note the percentage of documents that criticize teaching, curriculum, and instruction. The purpose of this is not to make correlations or predict patterns, but rather to quantify adjectives such as "most," "some," or "few." This approach is not uncommon in interpretive comparative analyses; see for example Elder's analysis of post-colonial Indian textbooks.[14]

Conceptual Framework

The analysis is guided by theory from intellectual history and cultural geography on the social construction of a cultural "other." In particular, this body of theory examines how a more powerful region (e.g., the West) creates an "other" in a less powerful part of the world (e.g., the Middle East, Eastern Europe) that reinforces the creator's sense of superiority. While the "other" is partially based on reality, it is also a selective and distorted construction. It is similar to a stereotype, "a standardized mental picture that is held in common by members of a group and that represents an oversimplified opinion, prejudiced attitude, or uncritical judgment."[15] Negative aspects are emphasized, while positive aspects that contradict the stereotype are ignored.

Constructions of the "other" have been examined in terms of individuals (male/female) or groups (white/black). Said extended theory about the social construction of the "other" by applying it to large parts of the world. In his study of attitudes toward the Middle East, he found that Western (i.e., American and European) scholars conceptualized the cultures and peoples of the region as backward, exotic, and childlike, a conception that both justified and reinforced colonialism and imperialism.[16] Orientalism as a field of study created an "other" to the "West" to act as a self-congratulatory mirror, and in so doing the "Orient" became all that the "West" preferred not to believe about itself. Thus, if the West saw itself as rational, the Orient was irrational. Likewise, the West saw itself as civilized, developed, orderly, productive; the Orient perforce was seen as uncivilized, undeveloped, disorderly, and unproductive.

By cloaking the Orient in such a way, the West was able to reinforce its positive self-image simply by "observing" or studying the Orient. Observation and study become severely limited, however, when the point of the exercise was to reinforce one's superior self-conception. Rather than seeing reality in its complex entirety, many Orientalists (i.e., scholars of the Middle East) limited their observations to what fit their conception. Like a stereotype, Orientalism saw what it wanted to see and ignored that which was contradictory.

Following on Said's insights, scholars of intellectual history and cultural geography have recently examined how Europe has been conceptually divided into "East" and "West" from the Enlightenment up to the present.[17] This division is based on concepts rather than physical geography, and is thus a social and cultural construction. For example, Prague lies geographically west of Vienna, even though many people would consider Austria a member of the "West" and the Czech Republic a member of the "East." As a social construction, "Eastern" and "Western" carry cultural connotations. Historically Western Europe has been constructed as enlightened, civilized, and cultured, while Eastern Europe is backward, barbaric, despotic, primitive, and exotic.[18]

Even after the end of the Cold War, the divide between Eastern and Western Europe remains, despite many shared cultural values and traditions. Moreover, there are now additional connotations: the West as tolerant, efficient, active, developed, organized, and democratic, and the East as intolerant, corrupt, passive, undeveloped, chaotic, and undemocratic.[19] More benevolent yet still negative connotations claim "Eastern Europe" is in transition to becoming modern and democratic; in other words, it is not yet fully modern or democratic. A trivial but telling example can be seen on the Lonely Planet travel guide's website for the Czech Republic. Under the country name is the subtitle, "A fairytale land rushing headlong to modernity."[20] While the author of the text is plausibly using a looser definition of modernity than the standard academic one, it is unlikely that the same claim would be made for Italy, Austria or any other country of the "West."

One way that socially dividing constructions can exist is through the deliberate selection of attributes to maintain stereotypes. Social constructions of the "other" are powerful because they are based on qualities that to a certain extent are true. The problem, however, is that in order for the construction to be maintained as an "other," qualities in the society that would negate the construction's connotations must be ignored. Social constructions of the "other" are based on selective, and thus distorted, conceptions.

Findings

Sixty-eight percent (65 of 96) of all documents included in this study have as their topic educational change or reform in the post-communist countries. All but a few of these documents are descriptive accounts of change, rather than empirical or analytical studies examining the reasons, rationales, and processes of educational change and reconstruction. Eighty-nine percent (58 of 65) of the documents about educational change negatively evaluate either ongoing reforms or the current state of post-communist schooling, with most authors making recommendations how they believe post-communist schooling specifically should change.

The topic of 29 percent (28 of 96) of all documents in the sample concerns collaborative projects between American and post-communist partners. Eighty-two percent (23 of 28) of these documents do not mention any learning by the American participants. Rather, the overwhelming focus of this group of documents is on what the American participants taught, and what the post-communist partners learned. These documents typically describe workshops, grants, and exchanges whose purpose is to transmit knowledge and expertise from the West to the post-communist countries. I therefore conclude that the authors perceive these projects as a one-sided transmission of knowledge and skills from American experts to their needy post-communist colleagues, rather than a truly collaborative project that would lead to learning for both parties.

A strikingly small proportion, six percent (6 of 96), of the documents are about American efforts to directly learn about "best practices" in the post-communist countries, especially Russia. They include studies of reading instruction, educational administration and leadership, foreign language learning, and teacher satisfaction. That this group of studies is markedly small shows that the great majority of American scholars involved in projects in the region viewed their work as reconstruction of post-communist education rather than mutual learning between peers.

A larger minority of documents notes any positive aspects of post-communist schooling. Overall, 25 percent (24 of 96) of the documents note a positive feature, mostly about teaching and learning. Fifty percent of the documents noting a positive feature (12 of 24) single out aspects related to teaching (e.g., teacher training and quality, curriculum, instructional techniques). Fifty-four percent (13 of 24) mention learning, especially in terms of student motivation, enthusiasm, dedication, and ability.

While one-quarter of the documents acknowledges positive aspects of post-communist education, a much larger group is primarily negative in its evaluation of schooling in the region. Sixty-nine percent (66 of 96) of the documents have a negative overall evaluation of their subject; the remaining 31 percent either have a neutral or positive evaluation. A document was considered to be negative if it only mentioned negative aspects. By contrast, a neutral document withheld any kind of evaluation, or acknowledged both positive and negative aspects. In most of the negative documents, the author's stance that education in the region needed to change was explicit.

Of the negative documents, 73 percent (48 of 66) criticized aspects related to pedagogy, including curriculum, instruction, and teacher training and quality. Particular criticisms are that teachers focus too much on memorization and facts and not enough on application, problem-solving, reasoning, analysis, and "critical thinking"; teachers are too controlling and authoritative; classrooms should be centered more around the student than the teacher; and teachers are passive, inflexible, and unable to adapt or take initiative. The image painted by many of the American scholars is that the typical teacher in a post-communist country is traditional, outmoded, unprogressive, authoritarian, unable or unwilling to change, unprofessional, and ineffective. While the overwhelming target of negative judgments is related to teaching, *none* of the negative documents criticized students or student outcomes. None of the studies calling for reconstruction of post-communist education based their claims on a perceived or empirically proven deficiency among students. Rather, students were the most common object of American praise. This contradiction between deficient pedagogy and excellent learning was not noted by any of the scholars included in the study.

Most of the texts included in this study were written by academics without expertise in the region. Instead of being experts of Russian or Polish education, for example, they are typically experts of other subfields of education, such as teacher training, pedagogy or educational psychology. A few of the authors included in the study, however, could be classified as area specialists. While these area specialists tend to provide more sophisticated and knowledgeable analyses, they are not immune to many of the negative biases and stereotypes found among the other academics. This is not surprising, however. As Said documents in *Orientalism*, the leaders of the Western construction of the Near East as an exotic and inferior Other were Orientalists, scholars of the region.

Education for Democracy

A significant *leitmotif* underlying many of the documents in this study is that post-communist schooling is undemocratic. Forty-one percent (39 of 96) of all documents connect educational reconstruction in the post-communist countries with democracy or democratization. An even larger share of the negative documents—61 percent (37 of 66)—makes an explicit reference to democracy. Even more striking, 92 percent (36 of 39) of all documents that mention democracy have a negative orientation toward post-communist schooling.

Many authors either explicitly state or implicitly assume that the ultimate purpose of changing schooling practices in the post-communist countries is to foster democratization in the larger society. Indeed, improving the quality or quantity of education in general receives little attention. Instead, the most common rationale for reconstructing post-communist schooling is to make it and the larger society more democratic. For example, an academic who worked in Bulgaria and the Czech Republic remarks:

> The development of a new philosophy of education is the most important problem facing the former Eastern bloc countries.... Subsequently, any new philosophy of education requires the development of new methodologies of education—those that will help students develop the knowledge, values, and skills required to meet the challenges of a democratic society.[21]

Similarly, an academic who taught in Hungary asserts that schools in the post-communist countries are not democratic:

> At present, the schools, though generally successful in academic terms, do not really reflect the larger culture or the goals of democratization in the region.... If they [schools] continue to act only as relatively disconnected places where subject matter is transmitted to those who want to receive it, then education for democratic citizenship and participation in the economy will have to happen elsewhere.[22]

Implicit in these claims is the assumption that the forms of pedagogy common in the post-communist countries are undemocratic or unable to foster democratic attitudes, values, skills, and behaviors in students. In particular, it is commonly argued that for democracy's sake, schools must adopt active learning methods:

> One important way that educators can prepare students for participating in a democratic society is through the use of active methods of teaching and learning. The challenge to every teacher... is to create an environment in which students are encouraged to think critically and interact with subject-matter, peers, and teachers in ways that promote democratic behaviours and attitudes as well as mastery of academic content. Students who are asked only to be passive recipients of knowledge will not develop the skills necessary for engaging in the public discourse so essential to a successful democracy.[23]

Active learning methods are presumed to develop active citizenship, whereas traditional, teacher-centered methods are seen to develop passivity.

In addition to a lack of active learning, it is commonly argued that schools do not foster tolerance, respect, or the ability to express one's opinion.

> Clearly, there is much conflict and confusion within the Russian educational system at the moment. . . . Perhaps most critical is the hard task of nurturing democratic habits of thought and action . . . open debate, respect for opponents' positions, and the ability to judge a person on merit do not come easily.[24]

Also argued to be lacking are problem-solving and independent thinking skills:

> Problem-solving is a universally acknowledged, professional, curricular ingredient. Its absence sends the wrong message to the public about what constitutes a responsible citizen. . . . Although the school system in the Eastern countries taught everyone to read and to calculate, by conscious design, it tried to not teach students to think for themselves. Now, what can be done? The challenge facing the Eastern countries is to shift the emphasis in their pedagogical purposes upwards—from factual recall to higher order skills.[25]

Implicit in these claims is the assumption that (a) post-communist schools do not teach students how to think, express their opinions, or be tolerant of others, and therefore (b) they are undemocratic. They will become democratic only when they adopt the goals and processes of instruction that American academics take for granted as essential. For example, two professors of education sent to Russia to conduct a workshop on "teaching in a democratic society" note:

> We talked about the fact that the ability to think, solve problems and express unique thoughts and feelings was critical to participation in democracy. . . . Did the Ukrainian teachers learn to teach children how to participate in a democracy? Who can say? On a return trip, we witnessed the teachers implementing principles of democratic teaching that were introduced at the workshop.[26]

The authors argue that Ukrainian teachers have to be taught how to develop democratic skills in students. The assumption is that they were not doing this until the American workshop leaders taught them how.

Discussion

It is true that pedagogy in most post-communist schools is more traditional, frontal and teacher-centered than the more progressive and individualized techniques common in the United States and other English speaking countries. It is also true that post-communist schools put more emphasis on knowledge and concepts, and less emphasis on discussion, application or problem-solving. Finally, active learning methods and classroom discussion can develop skills in students that may lead to increased civic knowledge and engagement.[27]

It is problematic, however, to assume that the absence of this style of instruction necessarily means that schools are inadequately preparing students for participation in a democratic society. Students could be learning critical thinking, creativity, tolerance, and respect, and problem-solving in other ways in school, or even through experiences outside of school. Indeed, none of the scholars who argue that post-communist schooling is undemocratic based their claims on empirical or even anecdotal evidence of presumed deficiencies in students. And as noted earlier, most of these countries had already achieved "full" democracy in terms of political and civil liberties almost immediately after the fall of communism. Thus, neither evidence about students nor the larger society supports the notion that post-communist schooling is undemocratic.

The largest cross-national study to date of civic engagement, knowledge, and attitudes found that students in the post-communist countries did not score remarkably different on any one dimension compared to other countries or regions.[28] For example, students from Chile, Colombia, Cyprus, Poland, Portugal, Romania, and the United States had the highest levels of civic engagement. The only area in which students from the post-communist countries scored significantly lower than students from other parts of the world concerned the degree of trust that students feel toward their government. Judith Torney-Purta and associates explain this as a legacy of communism, and other similar studies have found the same finding among adults.[29] Lower levels of trust toward government can be problematic, but there is no evidence that it is an indicator of undemocratic attitudes or behavior.

When discussing education and democracy in the post-communist countries, American scholars focus on pedagogical and school-level factors, such as the aims and methods of instruction, teaching styles, and educational philosophies. Macro-level structural factors that are important for democratic education, such as equality of opportunity and outcome, equitable resources and funding, or language rights for ethnic minorities, receive scant attention. In many of these policy areas, schooling in the post-communist countries is highly democratic. For example, there are no dramatic differences in school funding that privilege the middle and upper classes while disadvantaging education-ally underserved groups, such as are common in the United States.[30] Rigorous academic secondary education is available for free to all able and motivated students, regardless of their family income. A similar quality of education in the United States is available primarily to stu-dents from wealthy families who live in exclusive suburbs with highly resourced public schools or who can afford private education.

Thus, American scholars' arguments that post-communist school-ing is undemocratic are based on a distorted and limited conception of democratic education. It privileges the "progressive" and individ-ualized pedagogy common in the United States as more democratic, and maintains—with little evidence—that the "traditional" and whole group approach common in the post–communist countries is less dem-ocratic. The conception is based primarily and often exclusively on aspects that are considered negative, while arguably positive aspects are ignored. It also focuses on teaching and learning, with very little dis-cussion for educational structures, systems or policy issues.

The American scholars' overwhelming negative perceptions of post-communist schooling may derive from a preference for progressive and individualized pedagogy. This privileging of "progressive" over more traditional forms of pedagogy has a long history, and is certainly the dominant philosophy in most schools of education. Governmental and parental calls for greater emphasis on standards and the "basics" have been growing, but are derided by many if not most education faculties, teachers, and professional associations.[31]

American scholars' long-held preference for progressive pedagogy does not explain all of their negative perceptions about post-communist schooling, however. Many other countries also have traditional teacher-centered and whole class approaches to teaching, yet they are not crit-icized as inferior or undemocratic. For example, typical pedagogical approaches in many Asian countries could also be described as tra-ditional, collectivist, and frontal, yet they are rarely if ever singled out by American scholars as undemocratic or in need of reform. Moreover, very few of the articles about education in Asian countries are negative or judgmental in the way typical of the post-communist documents. On the contrary, much of the American literature about Asian education extols its strengths and successes.[32]

Clearly there are other factors explaining American scholars' largely negative perceptions of post-communist schooling. I argue that these scholars are influenced by a social construction of Eastern Europe that paints the region as backward and inferior to that in the "West," including the United States. Said showed how such constructions of the "other" are characterized by limited, distorted, and stereotypical views that emphasize negative aspects while ignoring positive ones.[33] Wolfe, Burgess, and Delanty showed how the social construction of Eastern Europe as an "other" to the "West" developed historically and continues to exist after the end of the Cold War. For the last several hundred years Western scholars have consistently conceptualized the culture and political and social institutions of "Eastern Europe" as inferior. I argue American scholars' depictions of post-communist schooling follow the same trajectory.

Conclusion

American educational reconstruction efforts since the end of the Cold War in post-communist Europe are based on the notion that schooling in the region needs to democratize. It is commonly argued that democratization of education will create not only better schools but also a more democratic society. While widespread, such arguments are not based on evidence about a lack of democratic transition in the larger society, or a lack of democratic attitudes or engagement among youth. American calls for educational democratization are just as commonly heard in reference to Russia, a country that has yet to be considered fully democratic since the end of the Cold War, as they are for the Czech Republic, a country that was first considered fully democratic just one year after the fall of communism. Likewise, the largest cross-national study to date of democratic dispositions among young people found no significant differences between students in the post-communist countries of Europe and students in other parts of the world about civic knowledge, engagement. and attitudes.[34]

American scholars who conceptualize post-communist schooling as undemocratic most likely do so reflexively rather than deliberately. A convenient way to avoid charges of ethnocentrism or cultural arrogance is to cloak criticisms of post-communist schooling under the mantle of a benign and positive concept such as democracy. By conceptualizing post-communist schooling as undemocratic, American scholars are able to maintain their sense of superiority without tarnishing their sense of fairness. Rather, they can feel comfortable that they are promoting a just and necessary cause.

The belief that post-communist schooling is undemocratic makes common sense to many Americans since the region was subjected to totalitarian government until the last decade of the twentieth century. According to this view, schooling in the post-communist countries is naturally undemocratic as the societies had only had democratic rule since the early 1990s. And as the winner of the Cold War, the United States is naturally considered a paragon of democracy. Ironically, American scholars use the concept of democracy to justify their undemocratic imposition of ideal schooling onto their colleagues in post-communist Europe, just as their colonial ancestors justified inhumane schooling practices under the benign notion of "civilizing" the native population.

Adam Burgess argues that Westerners use democracy as an imperialistic tool in many of their interactions with post-communist countries. He contends that democracy in the post-communist era is serving the same justification for imperialism as civilization did in the nineteenth century:

> Western institutions are now central to determining policy through the region....Interference is based upon the unsupported prejudice that the peoples of the region are prone to "reject democracy."...Such interference is regarded as benign, and there is certainly little protest from the likes of Albania or Bulgaria who, because of their own marginalisation, have proven glad of any interest—on virtually any terms....But it is important to recall that even classical nineteenth century imperialism was inseparable from crusading morality. It was bringing education, Christianity and the like to the unfortunates. And no doubt many, such as the missionaries, fervently believed in what we now recognise to have been little more than self-congratulatory propaganda. Simply because Western interference declares that it is there to do good...does not mean that this is indeed beneficial, unladen and assures a beneficial outcome.[35]

As Burgess notes, efforts to serve others who are perceived to be in need are often really about reinforcing one's sense of superiority. By using democracy to justify such service, this motive can remain hidden.

While some educators in the post-communist countries have experienced fruitful collaboration with and learning from American

colleagues, others have also noted arrogance and an inability to understand contextual differences.[36] Overall, it is likely that educators in
the post-communist countries have ambivalent experiences about
American aid, similar to that found by Jeanine Wedel in the fields of economics and public policy.[37] As Wedel discovered, the benefit of foreign
aid most consistently valued on the ground were the financial resources
that came along with it, such as photocopiers and travel accounts. In
the field of education there have been similar remarks.[38] By contrast,
it is likely that the American scholars who participated in educational
reconstruction efforts enjoyed many benefits, such as expense-paid
stays abroad, publications, and lucrative grants. Even more important, however, may have been the fulfillment and sense of purpose that
accompanies the role of an expert dispensing essential and hereto now
unfamiliar knowledge about such an important concept as democracy.

Notes

1. Ladislav Cerych, "Educational Reforms in Central and Eastern Europe: Processes and
   Outcomes [1]," *European Journal of Education* 32, no. 1 (1997): 75–96; Pavla Polechova,
   Jana Valkova, and Radmila Dostalova, "Civic Education for Democracy in the Czech
   Republic," *International Journal of Social Education* 12, no. 2 (1997): 73–83; Jana Svecova,
   "Czechoslovakia," in *Education in East Central Europe: Educational Changes after the Fall of
   Communism*, ed. Sjoerd Karsten and Dominique Majoor, *European Studies in Education* (New
   York: Waxmann Munster, 1994), 77–118.
2. Heyneman, S. P. (1991) Revolution in the East: the Educational Lessons, in *Reforming Education
   in a Changing World: International Perspectives*, Alexander, K. & Williams, V. (Eds.) (Oxford:
   Oxford International Roundtable on Educational Policy), 35–47. K. Roberts, "The New
   East European Model of Education, Training and Youth Employment," *Journal of Education
   and Work* 14, no. 3 (2001): 315–328; TIMSS International Study Center, "Highlights of
   Results from Timss," (Boston, MA: Boston College, 2000).
3. Cerych, "Educational Reforms in Central and Eastern Europe: Processes and Outcomes [1],"
   75–96.
4. Ibid.; K. Roberts et al., "Employment and Social Mobility: Evidence from Armenia, Georgia
   and Ukraine in the 1990s," *European Journal of Education* 35, no. 1 (2000): 125–136.
5. Anthony Jones, "The Educational Legacy of the Soviet Period," in *Education and Society in
   the New Russia*, ed. Anthony Jones (Armonk, NY: M. E. Sharpe, 1994); Ken Roberts
   and T. Szumlicz, "Education and School-to-Work Transitions in Post-Communist Poland,"
   British Journal of Education and Work 8, no. 3 (1995): 54–74.
6. Roberts and Szumlicz, "Education and School-to-Work Transitions in Post-Communist
   Poland," 54–74. Bernard Thomas Streitwieser, "Memory and Judgement: How East German
   Schools and Teachers Have Been Regarded in the Post-Unification Decade," in *Education in
   Germany since Unification*, ed. David Phillips, *Oxford Studies in Comparative Education* (Oxford:
   Symposium, 2000).
7. Fulbright Scholar Program, Council for International Exchange of Scholars (www.cies.org)
   (accessed April 15, 2007).
8. Cerych, "Educational Reforms in Central and Eastern Europe: Processes and Outcomes [1],"
   75–96.

9. http://coe.int/T/E/Com/About_COE/ (accessed May 7, 2007).
10. Freedom House, http://freedomhouse.org (accessed May 1, 2007).
11. CIA World Factbook, http://cia.gov/library/publications/the-world-factbook/ (accessed May 7, 2007).
12. TIMSS, http://nces.ed.gov/timss/ (accessed May 7, 2007).
13. Edward W. Said, *Orientalism* (New York: Random House, 1978).
14. Joseph Elder, "The Decolonization of Educational Culture: The Case of India," *Comparative Education Review* 15, no. 2 October (1971).
15. http://mw1.merriam-webster.com/dictionary (accessed May 7, 2007).
16. Said, *Orientalism*.
17. Adam Burgess, *Divided Europe: The New Domination of the East* (London: Pluto Press, 1997); Gerard Delanty, *Inventing Europe: Idea, Identity, Reality* (New York: St. Martin's Press, 1995); Larry Wolff, *Inventing Eastern Europe: The Map of Civilization on the Mind of the Enlightenment* (Stanford, CA: Stanford University Press, 1994).
18. Wolff, *Inventing Eastern Europe: The Map of Civilization on the Mind of the Enlightenment*.
19. Burgess, *Divided Europe: The New Domination of the East*; Delanty, *Inventing Europe: Idea, Identity, Reality*.
20. http://lonelyplanet.com/worldguide/destinations/europe/czech-republic/ (accessed May 7, 2007).
21. Deborah M. De Simone, "Educational Challenges Facing Eastern Europe," *Social Education* 60, no. 2 (1996): 104.
22. Rick A. Breault, "Educational Reform in Eastern Europe: An Interview with Rick A. Breault," *The Educational Forum* 63, no. 3 (1999): 243.
23. Dawn M. Shinew and John M. Fischer, "Comparative Lessons for Democracy: An International Curriculum Development Project," *International Journal of Social Education* 12, no. 2 (1997): 115.
24. Stephen T. Kerr, "Diversification in Russian Education," in *Education and Society in the New Russia*, ed. Anthony Jones (Armonk, NY: M. E. Sharpe, 1994), 69–70.
25. Heyneman, "Revolution in the East: The Educational Lessons," 37.
26. Carol Seefeldt and Alice Galper, "Lessons from Ukraine," *Childhood Education* 74, no. 3 (1998):139, 142.
27. Judith Torney-Purta et al., *Citizenship and Education in Twenty-Eight Countries: Civic Knowledge and Engagement at Age Fourteen: Executive Summary* (Amsterdam: IEA, 2001).
28. Ibid.
29. Richard Rose, "Postcommunism and the Problem of Trust," in *The Global Resurgence of Democracy*, ed. Larry Diamond and Marc F. Plattner (Baltimore, MD: The Johns Hopkins University Press, 1996).
30. In the United States, for example, per student funding can vary up to 400 percent. See Bruce Biddle and David Berliner, "What Research Says About Unequal Funding for Schools in America," in *Education Policy Reports Project (EPRP)* (Tempe, Arizona: Arizona State University, 2002), 1–23; Jonathan Kozol, *Savage Inequalities: Children in America's Schools* (New York: Harper Collins, 1991).
31. See David Labaree, *The Trouble with Ed Schools* (New Haven, CT: Yale University Press, 2004); E. D. Hirsch, *The Schools We Need: And Why We Don't Have Them* (New York: Anchor, 1997); Diane Ravitch, *Left Back: A Century of Battles over School Reform* (New York: Simon and Schuster, 2000).
32. Kai-ming Cheng and Kam-cheung Wong, "School effectiveness in East Asia: Concepts, Origins and Implications," *Journal of Educational Administration* 34, no. 5 (1996): 32–49; David Reynolds, *World Class Schools: International Perspectives on School Effectiveness* (London: Routledge, 2002); Harold W. Stevenson and James W. Stigler, *The Learning Gap: Why Our Schools Are Failing and What We Can Learn from Japanese and Chinese Education* (New York: Simon & Schuster, 1994).

33. aid, *Orientalism*; Wolff, *Inventing Eastern Europe: The Map of Civilization on the Mind of the Enlightenment.*
34. Torney-Purta et al., *Citizenship and Education in Twenty-Eight Countries.*
35. Burgess, *Divided Europe: The New Domination of the East*, 110–111.
36. Ladislav Cerych, "Educational Reforms in Central and Eastern Europe," *European Journal of Education* 30, no. 4 (1995): 75–96; Polechova, Valkova, and Dostalova, "Civic Education for Democracy in the Czech Republic."
37. Jeanine R. Wedel, *Collision and Collusion: The Strange Case of Western Aid to Eastern Europe 1989–1998* (New York: St. Martin's Press, 1998).
38. Joy Morrison, "The Changing Model of Russian Media and Journalism Education," *Journalism & Mass Communication Educator* 52, no. 3 (1997).

Bibliography

Burgess, Adam. *Divided Europe: The New Domination of the East.* London: Pluto Press, 1997.

Delanty, Gerard. *Inventing Europe: Idea, Identity, Reality.* New York: St. Martin's Press, 1995.

Elder, Joseph. "The Decolonization of Educational Culture: The Case of India." *Comparative Education Review* 15, no. 2 (October 1971): 288–295.

Havel, Václav. *The Art of the Impossible: Politics as Morality in Practice.* Trans. Paul Wilson. New York: Alfred A. Knopf, 1997.

Huntington, Samuel. *The Clash of Civilizations and the Remaking of World Order.* New York: Simon and Schuster, 1996.

Klaus, Václav. *Renaissance: The Rebirth of Liberty in the Heart of Europe.* Washington, DC: Cato Institute, 1997.

Kozol, Jonathan. *Savage Inequalities: Children in America's Schools.* New York: Harper Collins, 1991.

Ravitch, Diane. *Left Back: A Century of Battles over School Reform.* New York: Simon and Schuster, 2000.

Said, Edward W. *Orientalism.* New York: Random House, 1978.

Seefeldt, Carol, and Alice Galper. "Lessons from Ukraine." *Childhood Education* 74, no. 3 (1998): 136–142.

Streitwieser, Bernard Thomas. "Memory and Judgement: How East German Schools and Teachers Have Been Regarded in the Post-Unification Decade," in *Education in Germany since Unification*, ed. David Phillips, 57–80. Oxford: Symposium, 2000.

Torney-Purta, Judith, Rainer Lehmann, Hans Oswald, and Wolfram Schulz. *Citizenship and Education in Twenty-Eight Countries: Civic Knowledge and Engagement at Age Fourteen: Executive Summary.* Amsterdam: IEA, 2001.

Wedel, Jeanine R. *Collision and Collusion: The Strange Case of Western Aid to Eastern Europe 1989–1998.* New York: St. Martin's Press, 1998.

Wolff, Larry. *Inventing Eastern Europe: The Map of Civilization on the Mind of the Enlightenment.* Stanford, CA: Stanford University Press, 1994.

CHAPTER NINE

Lost in Translation: Parent Teacher Associations and Reconstruction in Bosnia in the Late 1990s

DANA BURDE

International donors provided scant funding for education in the Balkans during the war in the 1990s and during reconstruction immediately afterward. Instead, both international and domestic aid stressed civil society building projects. The U.S. Agency for International Development (USAID) gave priority to programs that were intended to promote peace and reconciliation and to initiatives that might hasten the development of democratic institutions. Aid agencies considered civil society and community building projects critical to these efforts.

Despite the lack of resources specifically targeting education, some international nongovernment organizations (NGOs) succeeded in garnering support for education programs by billing their education projects as community building work. Developing Parent-Teacher Associations (PTAs) was the key element in several of these programs. Similar to the "community school" model frequently used in traditional development work elsewhere, in Bosnia these PTAs created and supported preschools, and were meant to assist with finding classroom space and teacher financing, and supporting children's learning.

The U.S. PTA model resonated with U.S. donors and program implementers; it also represented one of the key exemplars cited by leading social capital theorist Robert Putnam. According to Putnam, PTAs were ideal types of civic associations, essential in their role as building blocks for civil society. The combined focus of USAID on civic engagement, with the simultaneous rise of Putnam's theories gave added impetus to the perceived power of the PTA to transform communities. This chapter describes the transformation of the PTA model as it moved from suburban American schools to post-conflict Bosnia. It argues that although PTAs are remarkable civic institutions and may hold promise in the future, they lacked the political capital necessary in postwar Bosnia to build strong social ties with the potential to reconcile fragmented communities.

The data was collected for this study in 1999 and 2000, shortly after the war in Bosnia ended. The study focused on an early childhood development program that was started and supported by an international NGO during and just after the conflict. The program was intended to provide early childhood development and safe spaces to children, but it was also intended to provide civic training to parents through PTAs and to stabilize communities in the process. The following pages provide a historical reflection on the period of reconstruction after the war in Bosnia, when the international aid community was focused on reconstruction that relied heavily on a particular understanding of social capital and civil society.

Social Capital Theory, Civil Society, and U.S. Aid to Bosnia

Social capital entered the lexicon of sociologists in the 1970s; James Coleman and Pierre Bourdieu developed it in the 1980s and early 1990s, and Robert Putnam is widely viewed as responsible for popularizing it several years later. This followed on the heels of the theoretical development of human and cultural capital, and is related to, but distinct from these predecessors. Bourdieu underscores its utility in economic terms: "Social capital is the aggregate of the actual or potential resources which are linked to possession of a durable network of more or less institutionalized relationships of mutual acquaintance and recognition—or in other words, to membership in a group."[1] The group provides its members with a type of social credit that is rooted in the network; an individual cannot carry it away from the group. If the individual leaves the group, the particular social capital that the group provided is no longer available to the individual. Robert Putnam defines social capital as "features of social organization, such as trust, norms and networks that can improve the efficiency of society by facilitating coordinated actions."[2] He adds to this definition: "for *mutual benefit*" and holds that "networks of organized reciprocity and civic solidarity" are a precondition for socioeconomic modernization.[3] Thus, according to this scholarship, successful governance is linked directly to levels of civic engagement.

In his 20 year study, *Making Democracy Work*, Putnam and colleagues described and assessed the success or failure of local government in northern and southern Italy.[4] The study was a landmark work among theorists of civil society and social capital for several reasons. First, it found that the more "civic minded" Italians in the north were able to create and demand more democratic institutions than their "uncivic" counterparts in the south because of greater reserves of social capital. Second, the combination of *what* the study said (that democracy could be strengthened by building civil society via building social capital) and *how* it said it (with simple methodological analysis), gave the study prominence in the world of international development. Although widely criticized for attributing causal importance to phenomena and historical summaries that were weakly associated with his findings in Italy, Putnam has continued to shape academic studies of civil society over the years with other seminal work such as *Bowling Alone* that possesses a similar attractiveness for its elegant and apparently straightforward argument.[5]

Despite these efforts, a theoretical overview of the concept of civil society breaks down quickly into multiple pieces. Scholars debate these various meanings, picking and choosing among the shards to reassemble sets of them again into a coherent whole. The theoretical dilemma lies in the fact that the different, opposing definitions of civil society are so closely related and overlap each other, in addition to being mistaken for their component parts (social capital, civic associations), that their analytical value seems to disappear. Michael Foley and Bob Edwards capture this spirit, noting that "the concept seems to take on the property of a gas, expanding or contracting to fit the analytical space afforded it by each historical or sociopolitical setting."[6]

The intense debates among theoreticians present a striking contrast to the way the term is implemented in daily life in the world of international development. In addition to the other definitional difficulties and labels they must cope with, nongovernmental organizations are also referred to as *constituting* civil society. Particularly since the early 1990s, many donor and implementing agencies have conflated the term *civil society* with NGOs. During this period of aid intervention in Bosnia, USAID for example, defined civil society as "nonstate organizations that can act as a catalyst for democratic reform."[7] This added confusion to the discussion. Since PTAs fostered by outside agencies were not generally registered as NGOs in countries receiving development assistance, they were excluded from external funding that they might have had access to otherwise, under another definition of civil society.

International nongovernmental organizations (INGOs) typically rely on their own definitions without explicitly addressing the political nature of civil society. Among international agencies working in Eastern Europe, Russia, or central Asia, different agencies or even different employees within the same agency rarely share a common definition of civil society. It can refer to "the organization of a society on the basis of a certain framework of civic values," implying a civil society that is complementary to a respectful, democratic state; or sometimes, "to a particular set or range of institutions, organizations and movements within a society," implying that the existence of organizations and actors themselves constitutes civil society.[8]

The INGO[9] whose Bosnia PTAs are examined here relies on a definition of civil society that values "independent civic associations" concerned with the provision of public services. It is assumed that they are independent because they are not managed or directly part of state bureaucracy. It is assumed that these associations are civic because they are occupied with managing and advocating for a service that is considered part of the general well-being of the community—early childhood education—a benefit for the community and described at times as a nonpolitical endeavor. As is often the case when civil society arguments are employed to support social service delivery, the political nature of these associations is not fully explained. These associations are meant to act politically to protect their interests—early childhood development—yet, within a paradigm of civil society that is complementary to the state, it is not clear how these organizations are supposed to do this. In addition, they are considered, by INGOs in general, and by the one studied here in particular, to have increased the density of civic associations (by their very existence), thus strengthening civil society.

The Parent Teacher Association as Panacea

PTAs are unusual and interesting organizations because they are expected to "double-task." They are expected to fulfill the purpose they were created to serve (improve local schools) and provide simultaneously a vehicle to produce add-on benefits that may relate to larger social issues (that in turn affect schools). Because they are a unique phenomenon, PTAs are frequently used to illustrate theoretical points.

The ways that PTAs figure in Coleman's notion of intentional organizations is particularly salient given how the translation of these structures to Bosnia eventually unfolded. In intentional organizations, members create an association from which they intend to benefit. PTAs are examples of these types of organizations. If a group of parents form a PTA for a school, this association constitutes social capital for the organizers, the students, the other parents of children attending the school and for the school itself. As the association functions, it creates two social capital byproducts. One is that the organization itself can serve as a vehicle for other purposes in addition to what it was originally intended to do. The second is the benefit of enjoying the public good that the PTA creates (i.e., improvements in school working hours or in the curriculum)—something that is available to all parents whether or not they participate in the association.[10]

In Putnam's work the PTA plays a small but symbolic role. He asserts that it has been "an especially important form of civic engagement in twentieth century America because parental involvement in the educational process represents a particularly productive form of social capital."[11] In the United States the PTA, as a membership organization, has declined nationwide from what it was in the 1960s. At its height, it encompassed nearly half of the parents in the United States. Putnam attributes its success to the fact that the "form of connectedness" it offers appealed to many Americans. And it was valuable for building social capital. In his words, "it is easy in our cynical era to sneer at cookies, cider and small talk, but membership in the PTA betokened a commitment to participate in a practical, child-focused form of community life."[12]

This observation addresses the heart of the paradox regarding the purpose PTAs serve and the kind of social capital they build. In the consensus model of civil society and social capital, PTA activity focuses on neighborhood discussions of the welfare of children, cookies, and small talk, and not necessarily in that order. Yet perhaps these neighborhood discussions are the very reason for the decline in PTA membership. If potential members feel the organization is largely irrelevant and impotent, they may not be motivated to join.

It is worth noting that Putnam and others describe social capital as consisting of dense "nonpolitical" social ties. For Putnam, the most exemplary forms of social capital were historically nurtured by women in female-dominated organizations that he views as nonpolitical. In fact, many scholars disagree with this characterization. Theda Skocpol and Elizabeth Clemens have researched decades of women's civic engagement in America and find that women's groups, including PTAs, were in fact, intensely political organizations. They find this was the case at the same historical moment Putnam uses to refer to PTAs as making valuable social capital contributions. In advocating for community interests, they shaped local, state, and national legislation.[13]

Despite ambiguity in the U.S. academic literature about how to characterize PTAs as a social institution, it remains commonplace to find them held up as an example of civic engagement par excellence. That the PTA both achieves educational improvements (in theory, at least) and has "spillover effects" on building social capital in broad terms were key factors in bringing them to war-ravaged Bosnia.

International Aid and the Bosnian Conflict

The war in Bosnia is considered one of the most brutal in modern European history by virtually all who have studied it, reported about it, or lived through it. The term *"ethnic cleansing"* although it describes similar practices in earlier European wars, gained common currency during the 1990s from this war, and refers to the intention of one group to create its own homogeneous national territory and rid itself of ethnic minorities. Hundreds of thousands of civilians were expelled from their homes and fled to other parts of the country or abroad. Tens of thousands of civilians were massacred or murdered in their homes. Few towns or cities in Bosnia exist that were not touched by the war. According to UNHCR, at one point the agency was helping three and a half million war victims in the former Yugoslav republics—refugees, the internally displaced and others.[14] Peace negotiations proceeded through fits and starts and false promises until mid-1995, when the United States and Europe attempted to revive the peace process. The "General Framework Agreement for Peace," or as it is more commonly referred to, the "Dayton Agreement," was reached in November 1995. For the following years, peace was achieved by maintaining a substantial international military and civilian presence and by creating an office for a special United Nations representative to regulate the government at all levels.

According to international NGOs, the war in the Balkans epitomizes many contemporary armed conflicts. In recent decades, 85 percent of conflict-related deaths were civilian, a rate that rose from approximately 50 percent from wars that were fought in the early part of the century.[15] Wars today occur generally within the border of one state, between groups of a different region or ethnicity.[16] They are protracted and produce outflows of refugees, as well as huge numbers of internally displaced people and civilians living in the midst of sporadic and unpredictable strife. INGOs distribute food and medical supplies as short-term emergency measures to accommodate basic needs of these populations. By providing education programs in Bosnia-Herzegovina, some international agencies tried to adapt their responses to these conditions, intending to have a larger social impact and a lasting effect.

For several decades conventional wisdom regarding international aid has divided assistance into two categories: emergency relief for regions facing natural disasters or war, and development for countries on the road to greater industrialization. Under this rubric, certain activities have been classified as "relief," such as distributing food and medical supplies, and others as "development," such as training and education. This distinction was founded on a belief in a linear progression of development: countries were thought to move from relief, to rehabilitation and finally into the development stage. The timing of the program studied here corresponded to many changes in humanitarian intervention. Goals became conflated: traditional emergency organizations began focusing on development work, and traditional development organizations began carrying out emergency work.

Early childhood education is seen as an important site for foreign intervention particularly in conflict-affected countries because of the intrinsic value it offers many parents and because of its perceived neutrality. Throughout the world, parents are often motivated to improve their children's opportunities in life by providing them access to education. In Bosnia-Herzegovina, the population is highly educated and the value of preschool was widely accepted before and after the war. As a social entry point for INGOs working in a politically charged environment, preschool has the added appeal of appearing apolitical. The innocence of small children seems to extend, by association, to programs designed for them.

Since this chapter discusses the responses of local actors to international interventions, it is important to note the educational conditions that preceded the war. Although the civil conflict was similar, Bosnia differs from many countries in which international emergency aid and development agencies are accustomed to working. Yugoslavia had an extensive infrastructure with established social and government institutions prior to the conflict. The literacy rate was similar to that of other countries in Eastern Europe, and "the relative prosperity, freedom to travel and work abroad, and landscape of multicultural pluralism and contrasts that Yugoslavs enjoyed were the envy of eastern Europeans."[17] Within the education system, however, early childhood education prior to the war was largely an urban phenomenon and functioned mainly as daycare for children rather than for their educational benefit.

The government of the former Yugoslavia maintained a goal to provide comprehensive early childhood education from the late 1960s until the war. Preschool administration was organized under the Ministry of Social Welfare. It followed the Eastern European model of providing day-long care to children aged one through seven, complete with beds and meals, as part of a system to allow women to join the work force. These age groups were divided and provided care accordingly: nurseries served children under three years of age, and kindergartens catered to children aged three to six.[18] Although pre-war statistics vary, by most estimates, between 2 percent and 12 percent of the total population of children of preschool age in the former Yugoslavia attended preschool.[19] According to 1999 Council of Europe report, Bosnia and Herzegovina had the "least developed and most inequitably provided level of education in the former Yugoslavia, with only 6% of children attending pre-school... as opposed to 90% in the former Autonomous Province of Vojvodina [part of northern Serbia today]."[20]

Regardless of the uncertain figures and its uneven delivery, early childhood care and education is a familiar and important concept to most parents in the former Yugoslavia. Unlike Europe and the United States where mandatory schooling usually begins at age five or six, in Bosnia it began and still begins in most places at age seven. As a result, parents are eager and sometimes even desperate to enroll their children in preschool, to keep them off the street and to prepare them for primary school.

Traditionally, preschool activities and family involvement were separate from each other; there was cooperation, but not partnership. Parents were invited to special events and performances and they paid for food for their children during the time spent in the preschool. They communicated with teachers only in the mornings and afternoons when they dropped off or picked up their children. Teachers' responsibilities in relation to parents included organizing meetings with parents four times each year. Extending parental access to the preschool beyond these examples was complicated. There were strict rules regarding hygiene and usually parents needed permission to enter the premises.[21]

Although prewar preschool education methodology did not require parents and teachers to build a strong relationship that aimed to increase parental involvement, and with it its attending ripple effects cited in the United States, neither did it exclude parents altogether. As described above, parents and teachers communicated regularly, and parents attended meetings and participated in some preschool activities, something that must be kept in mind as we consider the INGO sponsored program described below.

The PTA Model and Early Childhood
Education in Bosnia

At the time when the INGO early childhood development program in Bosnia was designed in 1993, international agencies sponsored very few education programs for civilian populations living in conflict zones. Conceptually, the INGO used the well-organized logistics of an emergency relief program delivering food or medical supplies. With similar deployment, the INGO meant to "distribute" a social service using a food distribution model. Although this idea had been proposed before by agencies interested in addressing educational needs by using a "school-in-a-box"[22] approach, these were often supplements to other programming; they were not generally financed to reach a large scale or to include extensive community involvement. By applying the techniques used during emergencies, the INGO hoped to demonstrate that education could be promoted during conflict and such a model could be transferred and adapted within the region and beyond. Although this has emerged today as a common approach for education in emergencies programs, at the time it was revolutionary among humanitarian aid workers. Furthermore, despite the fact that it was quite common for bilateral aid agencies such as USAID to carry out their work indirectly via INGOs, involving local communities directly in their own relief was one of the unusual aspects of this program.

As a financing organization, USAID works with implementing organizations (e.g., international development nongovernmental organizations), yet remains a powerful force since their requests for proposals frequently set agendas for reforms. Nonetheless, the international development nongovernmental organizations (e.g., Catholic Relief Services, International Rescue Committee, Save the Children Federation, World Learning, and World Vision, to name a few) that apply for and win these grants have a more direct impact on educational

program results than does the funding organization, since they implement the educational reforms that are designated by the requests for
proposals. In applying for these grants, the INGOs usually use the same
discourse designated by the funder to signal that they understand the
parameters of the work involved. During implementation, the process
enters another stage as INGOs add their own perspectives to the reform
once the grant has been won from USAID.

The goals of the INGO programs that aim to transfer and promote
parental involvement in educational abroad, in the case study described
in this chapter, state that the broad principles of the education program
are meant to provide community members with "a nation-wide network" and "significant nation-wide capacity building."[23] In describing
the role of parent volunteers in preschools, the INGO said:

> Parent volunteers in the classroom assist the teacher in supervis-
> ing small groups and providing individual care and in the process
> acquire better parenting and communication skills. Parent/citizen
> involvement in and commitment to the operation of a local, non-
> governmental institution is a practical and necessary step in the
> development of civil society. This program provides a large num-
> ber of citizens with the opportunity and experience.[24]

Thus, the INGO education program emphasizes parents' education,
decision-making skills, and participation in local organizations that
will lead to a wider impact on national educational policy, and, following Putnam, on civil society.

The INGO staff promoted the preschool program described here in
the midst of war-torn Yugoslavia because they believed it would provide multiple services to communities in crisis and critical protection to
small children. Specifically, the aims of the designers were as follows:

- to provide high-quality, inexpensive early childhood education
  via a new and innovative program for the former Yugoslavia;
- to offer a return to normalcy that provided increased stability in a
  war-torn society;
- to provide income generation for women;
- to increase civic participation resulting in increased civil society;
- to translate community initiatives into education policy reform.[25]

In line with many development INGOs, the INGO featured in this
case study believed that community commitment was the cornerstone

to its preschool program's success. Community ownership and local participation were essential ingredients to any work that it conducted.

Thus, although the education goals were important, the program was also a civil society-building project that hinged on the "best practice" of creating parent-teacher associations. To this end, the INGO designed an elaborate management and training system to establish three levels of association (parental, municipal, cantonal) to support the preschools. The most basic level was called "parent support groups" and had a similar functional and operational definition to a local American PTA chapter. These nascent groups were given training in organizing, fund raising, and management, with the aim of increasing parents' participation, helping them to advocate and fund raise for themselves, and ultimately to sustain the preschools independently of the INGO. The international organization reasoned that bringing concerned parents and teachers together to care for and educate children, would create the proper conditions for community mobilization and for enduring social change.

Simultaneously, the community-building efforts appealed to donors: more than half of the funds received during the life of the program were secured because of explicit references to community development as an add-on benefit gained from parental involvement in the preschool program. Thus, the program was flexible enough to accommodate both USAID's and the INGO's interests.

The INGO employed local facilitators, mentors, and sustainability trainers to organize parents into parent support groups, to provide support to teachers, and to help local associations access funds. Motivated parents from the parent support groups were invited to attend a community organization training session run by INGO staff, which usually was held at the end of the nine months of material support to the preschool, in preparation for forming municipal level organizations (MLOs). The MLOs were meant to pursue and supply funds to PTAs in order to maintain the preschools. Members received training in techniques such as conducting needs assessments and proposal writing. There were 16 INGO-sponsored MLOs in BiH at the height of the program. Nine continued to function in one canton in the federation until 1999, with the help of the INGO team of three trainers who provided training, seminars, and networking support to these associations. Three other organizations existed in some capacity without INGO support, and 4 were defunct as of summer 1999.

In practice, in pursuit of finances, some MLO leaders used innovative strategies to support the program. For example, one received a van donated from Italy to drive children to and from preschools. However, this was the exception to their activities rather than the rule. Although all established working connections with international agencies for material donations, and used personal connections for local institutional support, institutional funding was not forthcoming. One MLO leader, for example, showed me several proposals she had written in response to donors' requests for applications. The objectives were in line with the donors' requests, budgets were reasonable, and the organizational structure was sufficient, but each proposal had been declined. Without INGO representatives to facilitate the process, the nascent local associations could not bridge the divide between the local activities and global funding. These MLOs did not remain active for long after they stopped receiving funds or training from the INGO.

Attempting to animate communities with short-term commitments and inconsistent interventions produced the intended discourse but did not produce the intended results. Given that the INGOs, are accountable ultimately to their donors, it is difficult for a service-delivery organization such as the INGO described here to operate between opposing political paradigms. Donor requirements can conflict with, and significantly hamper, an INGO's mission to create "lasting, positive changes."[26] Or, in other words, INGOs, are not always in the best position to support the political interests of their program beneficiaries.

In BiH, conflicts emerged between the INGO and USAID when the INGO was set to begin work in the Serb Republic (RS) and USAID tied conditions to its grant making work there.[27] If the INGO agreed to accept funds in this situation, it was meant to manage the expectations of the donor by actively enforcing the donor's political conditions. The INGO accepted the funds, although the decision subsequently created difficulties for the organization's credibility with the new communities in which it was planning to work.[28]

However, the U.S. government was not the only governing body that shaped the work of the INGO. Local government officials also requested that the INGO compromise on its long-term goal of social mobilization. Although sustainability and community development were critical issues for the INGO in representing the program to donors funding the program as well as for its internal standards that emphasized the importance of parental involvement, it was not critical or even important to local government officials. After the peace agreement was signed in 1995 and the government systems began to reassert themselves, the INGO revised its community approach to education in order to appeal to the local educators with whom it had to work and to secure funding and government space for the preschools. Aspects of the education reform program that clashed with government education philosophy were modified or eliminated. For example, in the federation, the INGO downplayed its use of paraprofessional teachers and reinforced the educational value of the program. In negotiating program administration in the RS, the program was inserted into government kindergartens, thereby generally eliminating problems with space, but also placing the program back into the hand of exsocialist, government workers. Neither PTAs, nor parental involvement of any kind, was a priority for the local government, and the INGO did not advocate for supporting them.

As a result, after a short time of implementing the preschool model with its strong, original emphasis on parental involvement, almost all traces of the best practice promoted by the INGO vanished into the vortex of government bureaucracy. From the perspective of local actors—government officials, teachers, and parents—ultimately there was little convergence between the international model and the locally implemented program.

Conflicting demands placed on INGOs in implementing their programs affect their ability to promote model reforms, but, at the same time, the internal structure of the INGO itself may compound this dilemma in the attempt to create sustainability. There is confusion among the donors and implementing agencies alike about the definition and importance of sustainability, particularly during or after an emergency. International donors use the word *sustainability* to describe the potential of their investment to be adopted by the local community, managed and for the most part funded by local nonprofit, for-profit, or government groups. In the case presented here, sustainability was central to the program design from the outset, and it was the evidence used to back the INGO claims of success at civil society building. Instead of creating sustainability, however, the INGO seemed to create layers of professionals, isolating local actors from the global polity.

The INGO tracked and judged sustainability on several levels. First, numbers were critical in assessing a sustainable program. Important indicators were, for example, the number of preschools that remained in existence after the foreign funding was withdrawn, the number of parent support groups established, and the number of teachers trained. Second, recording community action that asserted rights to services was considered an important representation of systematic change. In addition, cost was linked directly to sustainability. The program was meant to be inexpensive without sacrificing quality so that "community members will both want it and can afford to own and operate it."[29] Included in the support were the teachers' manual, classroom consumable kits, hygienic kits, teacher remuneration, mentoring, and seminar costs. According to the first field office director, "If a preschool is able to operate one month after [foreign] funding has ended, then [the INGO] considers it sustainable." The international organization limited the definition because, "we didn't want the responsibility of continuing to monitor something that we didn't have the time or resources to track."[30]

These criteria only satisfied the short-term educational goals of the program; they did not promote social mobilization or provide parents with direct access to shaping "the frames that orient actors, including states."[31] If the INGO preschool program aimed only to provide a service, it did so. If, on the other hand, the INGO aimed to reach its longterm mission of creating lasting change, the program did not. It did not leave behind an extensive network of community associations that supported and advocated for innovative preschools, and that introduced into the nationwide preschool system child-centered learning, shorter education programs, and learning through play. Instead, it trained and funded a layer of local professionals who worked between local communities and international institutions until the funds ended and they sought jobs elsewhere. Thus, after the funding for the program ceased, the links between parent members of associations and broader structures were severed.

Conclusion

Perhaps accomplishing the short-term goals that this U.S.-modeled program intended to achieve was enough; many parents remembered the program with appreciation and gratitude for the solace it brought them in the midst of a violent and terrifying time. Perhaps expecting PTAs to mend communities healing after conflict was unrealistic and even unnecessary. Yet these expectations troubled the parents and teachers interviewed for this study. The messianic importance that U.S. funders placed on promoting democratic institutions, and the reliance on a key feature from U.S. society to do so, followed by abrupt policy lapses, reversals, or perceived abandonment, reduced the credibility of not only the U.S. intervention in the Balkans, but also of the faith in the reforms that it promoted.

Although the program provided a valued service, when the promises that civic action in the PTA held out failed to materialize, for many teachers gratitude was replaced with cynicism. Their interest in civic action and broader political participation diminished. One of the teachers expressed this sentiment. When asked about her plans to vote in the upcoming local elections, she said,

> I have voted in every election until now but I won't vote anymore because it doesn't matter. Look at this preschool. Who has money? All the money has been given to political parties and they spend it on their campaigns and we have none. Things here are only getting worse. What has changed? I changed the color of my hair, that's what changed. I changed the color and [my colleague] cut hers. (Interview, 1999)

This is not a study in which it is possible to draw a causal relationship between the failed promises of the broader, community building goals of the PTA, and the despair and resulting choices to opt-out illustrated by teachers' and parents' responses to questions regarding civic participation. It is likely, however, that both the small and large everyday acts of erratic U.S. foreign policy create an atmosphere of failed promises and dashed hopes that contribute to small, but important civic decisions. These decisions lead to decreased participation in a system that does not appear to be effective at realizing local goals. From Bosnia, to Iraq, to Afghanistan, the failure to meet raised expectations has contributed to the decline in respect for U.S. values and a decline in hope in U.S. assistance. In Bosnia, the PTA was unfairly tasked with these expectations, and, as a result, was a casualty in the post-conflict community building process.

Notes

Excerpts above from Dana S. Burde, "International Borrowing and Lending" in Gita Steiner-Khamsi, ed., *The Global Politics of Educational Borrowing and Lending*, pp. 173–187 (New York: Teachers College Press, 2004) reprinted by permission of the publisher.

1. Pierre Bourdieu, "The Forms of Capital," in *Handbook of Theory and Research for the Sociology of Education,* ed. John G. Richardson (Westport, CT: Greenwood Publishing Group, 1986), 249.
2. Robert Putnam, *Making Democracy Work* (New York: Simon and Schuster, 1993), 167.
3. Robert Putnam, "Bowling Alone: America's Declining Social Capital," *Journal of Democracy* 6, no. 1 (1995): 67. [emphasis added]
4. Putnam, *Making Democracy Work.*
5. Putnam, "Bowling Alone."
6. Michael W. Foley and Bob Edwards, "The Paradox of Civil Society," *Journal of Democracy* 7, no. 3 (1996): 42.

204 *Dana Burde*

7. USAID, 1996, cited in Jude Howell and Jenny Pearce. "Civil Society: Technical Instrument or Social Force for Change?" In *New Roles and Relevance. Development NGOs and the Challenge of Change*, ed. David Lewis and Tina Wallace (Bloomfield, CT: Kumarian Press, 2000), 80.
8. Thomas Carothers, *Assessing Democracy Assistance: The Case of Romania* (Washington, DC: Carnegie Endowment for International Peace, 1996), 65.
9. The INGO remains anonymous here to preserve confidentiality, but also because its work was largely representative of many INGOs at the time.
10. James Coleman, *Foundations of Social Theory* (Cambridge, MA: First Harvard University Press, 1990).
11. Putnam, "Bowling Alone," 69.
12. Ibid., 56.
13. Theda Skocpol and Morris Fiorina, eds. *Civic Engagement in American Democracy* (Washington, DC: Brookings Institution, 1999).
14. This figure includes victims of the conflict in Croatia.
15. Ruth Sivard, quoted in Everett M. Ressler, *Children in War: A Guide to the Provision of Services* (New York: The United Nations Children's Fund, 1993).
16. UNHCR, "Summary Of Registered Returns Of Displaced Persons Within Bosnia and Herzegovina." http://unhcr.ba.
17. Susan Woodward, *Balkan Tragedy: Chaos and Dissolution after the Cold War* (Washington DC: The Brookings Institution, 1995), 1.
18. UNICEF, "Early Childhood Programmes In Bosnia And Herzegovina." Sarajevo. 1996.
19. Council of Europe, *Education in Bosnia and Herzegovina: Governance, Finance and Administration* (Strasbourg: World Bank, 1999); Konferencija o predskolskom odgoju i obrazovanju u BiH, 1999; Kreso, interview, Sarajevo Faculty of Philosophy, 1999.
20. Council of Europe, *Education in Bosnia and Herzegovina*, 35.
21. Local NGO Director, interview, 1999, 2000.
22. In the area of preschool educational alone, Step by Step, the Soros Foundation/Open Society Institute (OSI) model for early childhood development, based on U.S. innovations in preschool education (High/Scope, Head Start, etc.) "annually provides training to more than 23,000 teachers in 28 countries or territories, serving more than 500,000 children and their families." OSI, "Children and Youth Program." http://soros.org/netprog.html.
23. [INGO], *Community Development Through Educational Support Projects: Second Annual Report* (October 1, 1995–September 30, 1996), 4.
24. Ibid., 22.
25. Ibid.
26. [INGO], "Mission Statement." INGO website.
27. According to USAID representatives, conditions also were applied to the Muslin-Croat Federation; however, in the case of the INGO program, these conditions were not implemented in either entity until *after* the USAID grant had been terminated in the Federation.
28. Interview with former staff member, 1999. One INGO staff member eventually resigned as a result of this decision.
29. [INGO], *Community Development Through Educational Support Projects*,
30. Interview with former staff member, 1999.
31. John Boli and George Thomas, eds., *Constructing World Culture: International Nongovernmental Organizations Since 1875* (Stanford, CA: Stanford University Press, 1999), 15.

Bibliography

Boli, John and George Thomas, eds. *Constructing World Culture: International Nongovernmental Organizations Since 1875*. Stanford, CA: Stanford University Press, 1999.

Bourdieu, Pierre. "The Forms of Capital," in *Handbook of Theory and Research for the Sociology of Education*, ed. John G. Richardson, 241–258. Westport, CT: Greenwood Publishing Group, 1986.

Carothers, Thomas. *Assessing Democracy Assistance: The Case of Romania*. Washington, DC: Carnegie Endowment for International Peace, 1996.

Coleman, James. *Foundations of Social Theory*. Cambridge, MA: First Harvard University Press, 1990.

Foley, Michael W. and Bob Edwards. "The Paradox of Civil Society." *Journal of Democracy* 7, no. 3 (1996): 38–52.

Howell, Jude and Jenny Pearce. "Civil Society: Technical Instrument or Social Force for Change?" In *New Roles and Relevance. Development NGOs and the Challenge of Change*, ed. David Lewis and Tina Wallace, 75–88. Bloomfield, CT: Kumarian Press, 2000.

Putnam, Robert. *Making Democracy Work*. New York: Simon and Schuster, 1993.

———. "Bowling Alone: America's Declining Social Capital." *Journal of Democracy* 6, no. 1 (1995): 65–78.

Ressler, Everett M. *Children in War: A Guide to the Provision of Services*. New York: The United Nations Children's Fund, 1993.

Skocpol, Theda and Morris Fiorina, eds. *Civic Engagement in American Democracy*. Washington, DC: The Brookings Institution, 1999.

Woodward, Susan. *Balkan Tragedy: Chaos and Dissolution after the Cold War*. Washington, DC: The Brookings Institution, 1995.

CHAPTER TEN

Allah, America, and the Army: U.S. Involvement in South Asia and Pakistan's Education Policy

M. AYAZ NASEEM

Shahzada, the eunuch character in Jamil Dehlavi's 1992 film *The Immaculate Conception,* describes the survival of Pakistan to be dependent on three factors; Allah, America, and the army. It has since become widely believed that politics and policy making in Pakistan is indeed to a large extent affected by these three elements. In this paper I explore the impact of the collusion of these factors on the educational policy in Pakistan. Specifically, I explore the impact that the U.S. involvement in South and South-Central Asia has had on the way education in Pakistan has been shaped by the U.S. involvement and conversely how changes in Pakistan's education affect the U.S. geostrategic interests in the region.

Pakistan has been hailed as one of the most valuable U.S. allies a number of times recently and in the past. However, the alliance and its value have been and continue to be contingent upon the U.S. geostrategic interests in the South and South-Central Asian region. Over the past five decades when the contingent weight (geostrategic and political) of Pakistan was important for the United States, the former was awarded (rewarded with!) hefty amounts of military and developmental aid. As soon as the U.S. interest in the region waned the aid and assistance was yanked away. Resultantly, the relationship never developed and flourished into a long-term "alliance." Due to the ad hoc nature of the alliance most of the U.S. aid to Pakistan was for military purposes. Pakistan's social and educational development was never a priority (except perhaps in the post-9/11 phase of the relationship). For the present exploration and analysis I examine two periods of U.S. involvement in the South and South-Central Asian region: 1980 to 1988 and 2001 to present. These can be termed as the two "Afghan episodes" of the U.S. geostrategic policy and of the U.S.-Pakistan relationship. The first of the two periods coincides with the Soviet invasion (and occupation) of Afghanistan and the U.S. proxy war waged through Pakistan and the *Mujahideen*. The second phase started after the September 11 terrorist attacks and the U.S. War on Terror against the Taliban.

This chapter begins by providing a historical background on the U.S.-Pakistan relationship from the 1950s to the present. I then provide a brief overview of Pakistan's education system and then explore and analyze the impact that U.S. involvement in the region has had on educational policy in Pakistan. Three overarching points should be kept in mind. First, during the initial Afghan episode (1980–87) the United States was not directly concerned with the Pakistani educational policy and the amount of aid and assistance for this sector was miniscule and indirect. During this period the United States ignored the mushrooming of *madrassah*[1] education in Pakistan, a position that suited the war effort against the Soviets. Second, since 2001 the United States has been much more concerned about the educational system in Pakistan especially the *madrassahs*. Third, the changes in Pakistan's educational system, especially the militarization of the curricula and the texts and the exponential growth of *madrassahs*, as discussed below are seen to have direct bearing on the U.S. interests in the region, on the War on Terror and on the overall security and stability of Pakistan.

Historical Context

Pakistan, which emerged as an independent nation state on August 14, 1947, has a unique geostrategic profile. Geographically, it is situated in a region that has immense strategic importance for the United States. Pakistan's long-standing rival India borders it on the east. On the west and northwest, Pakistan borders Afghanistan and China. Prior to 1971, by virtue of having its eastern wing (now Bangladesh) located in Southeast Asia it was important in the Southeast Asian strategic theater. Before the break up of the Soviet Union it was separated from the communist giant by a ten-mile strip called the Wakhan Corridor. On the south, Pakistan borders on the all important oil routes of the Indian Ocean through the Arabian Sea. The Arabian Sea also gives Pakistan a strategic proximity to the Middle East (West Asia) and thus the strategic oil reserves.

It is from this geostrategic importance that Pakistan derives its political and strategic importance in general and for the U.S. policy makers in particular. For example, with India leaning toward Soviet Union from the very beginning due to the socialist orientations of its first Prime Minister Jawaharlal Nehru, the United States courted Pakistan as an ally in its containment policy aimed at the Soviet Union in South, Southeast, and West Asia. This period saw Pakistan becoming a formal ally of the United States and the West through Southeast Asia Treaty Organization (SEATO), Central Treaty Organization (CENTO a.k.a. Baghdad Pact) and the bilateral Mutual Defense Assistance Pact with the United States. Membership in these bilateral and multilateral alliances made Pakistan a key player in a U.S. containment policy inspired by the "domino theory" of John Foster Dulles.[2]

However, with the *rapprochement* with China and the process of *détente* with the Soviet Union in 1970s Pakistan's contingent weight in the U.S. security policy started to decline. The decade of 1970s saw Pakistan dismembered,[3] alone and abandoned by the United States on the pretext of human right violations. It slipped from being a "most valuable ally" to being a pariah. It was not until the fall of the Shah of Iran, Ayatollah Khomeini's Islamic revolution in Iran, and the Soviet invasion of Afghanistan that Pakistan once again became important in the U.S. security schemes. From 1980 to 1988 Pakistan emerged as the key frontline state and a bulwark in the U.S. proxy war against the Soviet Union. Massive amounts of aid (approximately US$7 billion) were pumped in by the United States; Pakistan's territory was used to train and fund the *Mujahideen*[4]; and more than three million Afghan refugees poured into Pakistan straining the already meager social and environmental resources. The military regime of General Zia-ul-Haq was deemed indispensable by the United States and was propped up despite its abysmal human rights record.

In this period, significant changes came about in Pakistan's educational system. These changes (the militarization of the educational discourse, the mushrooming of militant *madrassahs*) have had far reaching impact not only on the educational system in Pakistan but also on the politics and geostrategic policies in the region and beyond. With the 1987 Soviet withdrawal from Afghanistan, the U.S. interest in

Pakistan once again diminished. Washington revived its concern about Pakistan's refusal to sign the Non-Proliferation Treaty and once again sanctions were imposed.

With the horrendous tragedy of September 11, the U.S. interest in the region and in Pakistan was once again rekindled. This time, however, Pakistan was neither asked nor wooed to become a frontline state in the U.S. War on Terror. It was categorically told that it had no option but to become an ally or be prepared to be bombed into the Stone Age.[5] General Pervez Musharraf, the military dictator at the helm of affairs in Pakistan took an immediate U-turn from his policy of supporting the Taliban and became an ally in the War on Terror. He was rewarded with a massive aid package (roughly US$10 billion). As in the 1980s during the first Afghan episode, the lion's share of the U.S. aid is earmarked for funding the military effort. A miniscule part of these funds is earmarked for social development. However, this time the policy makers in Washington are more concerned about Pakistan's educational system, particularly the *madrassahs*.

An Overview of the Education System in Pakistan

At its birth in 1947, 85 percent of the Pakistani population was illiterate. In backward regions of the country the literacy rate was even lower, with rural women having virtually a zero literacy rate. Ever since, successive governments have declared the attainment of universal primary education as an important goal. Although considerable resources have been expended in creating new infrastructure and facilities in the last 50 years, the literacy rate in Pakistan nevertheless remains low. Two-thirds of the population and over 80 percent of rural women are still illiterate. More than a quarter of children between the ages of five and nine do not attend school. Currently, the literacy rate in Pakistan hovers around the 50 percent mark.

This represents an impressive increase in literacy levels over the last ten years. From 37.2 percent in 1993 it represents almost 15 percent increase especially when plotted against the astonishing population growth rate that has averaged around 2.33 percent a year for the last many years. The literacy rate for males is 54.81 while that of females is 32.02 percent.[6] The Ten Year Perspective Plan 2001–11 propsed increasing the literacy rate to 61 percent (male 71.5%:female 50.5%) by 2005–6, to 68 percent (male 77%:female 65%) by 2010–11, and to 86 percent (male 86%:female 86%) by 2015–16. These are, indeed, overly optimistic projections that the government aims at in the context of its Education For All (EFA) commitments. The chances for achieving these goals are over ambitious and at best suspect.[7] Education in Pakistan is organized along primary (years 1–5), secondary (years 6–10), higher secondary (college years 11–12), and post-secondary (university years 13–16+) stages. Other than these formal education institutions a large number of students are enrolled in *madrassahs* that are charity-based institutions for religious education—dealt with in greater detail below. In terms of institutions Pakistan has a multitiered system. On one level these tiers are visible in terms of the public-private dichotomy (or "partnership," as the development planners in Pakistan term it). However, there are also multiple tiers of public schools. The top tier includes urban-based public schools that impart a better quality education, have competitive entrance requirements and enjoy considerable prestige. The second tier is composed of the urban-based government schools that are funded and managed by the provincial governments. The standards of these schools vary from one place to another. While some are better managed and impart a better standard of education, others are not as good. On the third tier are government schools based in the semirural and rural areas of Pakistan. These institutions often lack qualified staff and adequate infrastructure. Some are reported not even to have proper buildings.

Pakistan maintains a state monopoly over the production of textbooks for the public sector schools. While education is constitutionally a provincial subject, it is placed on a list of concurrent subjects[8] where the federal government provides the curricular direction. The prescribed textbooks are developed based on curricula approved by the curriculum wing of Pakistan's Federal Ministry of Education. These books are often badly designed and badly produced by approved publishers who have no input in the content of the textbooks.

United States and the First Afghan Episode: Islamization of Education in Pakistan

The Soviet decision to use force in Afghanistan in 1980 presented different options and opportunities to different regional and global players and had far reaching consequences for world politics and security in general. For an economically beleaguered Soviet Union, the Afghan adventure proved to be the proverbial last straw and was critical in the collapse of the Soviet state and its empire. For the Afghans, the Soviet invasion meant war, death, destruction, displacement, and a trans-formation of the society with deep and new conflict and fault lines. For the United States, the Soviet action presented an opportunity to avenge the Vietnam ignominy by bleeding its communist rival just as the Soviets had bled the United States in Vietnam. It was what Fred Halliday has called the advent of the second Cold War.[9] For Pakistan, the Soviet invasion of Afghanistan ushered in farreaching changes. Overnight Pakistan was turned into a frontline state. It also saw a mass exodus of refugees and displaced persons from Afghanistan to Pakistan, reaching a record figure of 3.2 million refugees in a matter of years. The Soviet invasion of Afghanistan also gave a lease of life for the mili-tary dictatorship of General Zia-ul-Haq. With the massive U.S. aid and an active political and diplomatic support the beleaguered junta was allowed to reign indefinitely without any pressure from the outside to democratize. The domestic demands for democracy were crushed with ruthless force without any fear of human rights oversight by the foreign governments or media.

With Pakistan becoming important for the U.S. strategic and secu-rity interests, once again large amounts of aid started flowing to and through Pakistan to fight the Soviet army in Afghanistan. It is estimated that the Zia government received close to US$7.2 billion in direct mil-itary aid. A large part of these funds were to bolster Pakistan military capabilities and were thus spent on buying military hardware from the U.S. defense industry. It is estimated that another US$6 billion were siphoned through Pakistan to the Afghan *Mujahideen* to enable them to wage jihad against the Soviet armed forces. A part of the funds was also used to buy loyalties of the religio-political elements that supported the military regime of General Zia or those who had influence among the Afghan commanders fighting the Soviets. Some of these religio-political leaders were later to set up establishments that trained and supported the Taliban.

In its zeal to make Afghanistan a Vietnam for the Soviets, the United States not only gave the military regime of General Zia the resources to suppress dissent and demands for democracy in Pakistan it also indi-rectly abetted the military regime's superficial process of Islamization. The so-called process of Islamization started by the Zia regime was directed less at invoking the social justice aspects of Islam and more at radicalization and militarization of the Pakistani society.

Washington ignored and tolerated Zia's Islamization as it served three key U.S. interests in the region. First, Islamization lent support to the military regime in Pakistan that was willing to fight a proxy war for the United States. Second, Islam could be positioned as an anticommunist ideology, which suited the U.S. war interests in Afghanistan. Third, the so-called Islamization mobilized a population in the name of the religion against the incursion of "godless" communist aggressor thus making it easier for the United States and its proxies in Islamabad to route the war effort through Pakistan. Zia's policy of Islamization specifically targeted three key institutions of the Pakistani society, namely the judiciary, the economy, and education. While much has been written about the first two,[10] here I will focus on the third.

Islamization of Education in Pakistan

The Islamization of knowledge and the educational system in Pakistan can be explored at many levels. However, in the interest of parsimony and due to the constraints of space let me briefly enumerate some of the salient ones. First, a number of institutes and commissions were formed and charged with the task of bringing the country's educational system into harmony with the principles and epistemological basis of Islam. Second, the regime patronized and appointed conservative elements as administrators and faculty in educational institutions across the country. Third, the regime proactively encouraged and promoted efforts at "Islamizing" sciences, anthropology, and economics. Fourth, it actively and overtly patronized students' wings of religio-political parties that supported the regime. Such patronage included funding and arming them as well as not bringing cases against them when they were responsible for violence. Fifth, there was a large scale Islamization of the curricula and the textbooks to emphasize the military aspects of Islam. As I have argued elsewhere, the militarization of the curricula and the textbooks effectively militarized the basis on which citizenship could be accorded.[11] Finally, the military regime patronized the growth and radicalization of the *madrassah* in Pakistan. Some of these *madrassahs* went on to become, what Zahid Hussain calls the "nurseries for *Jihad.*"[12] While others aspect of Zia's Islamization, particularly policies that were discriminatory toward women and minorities did come under the scrutiny of scholars, human rights organizations, and the U.S. administrations,[13] the changes in the educational realm did not come under scholarly scrutiny at the time.

Zia's Islamization campaign predated the resurgence of American interest in Pakistan. He had convened a national education conference in October 1977 with the stated objective of soliciting advice on ways to "bring education in line with the people's faith and ideology."[14] The new education policy promulgated by the military regime lamented the previous (Bhutto era) policy on neglecting primary education to a point where 78 percent of the population above five years of age was illiterate. The new policy vowed to undertake

> A fundamental reordering of national priorities in favor of a primary education...a comprehensive approach towards primary education which would include not only the augmentation of physical facilities but also measures to reduce the dropout rates, improvement in the quality of teaching and better supervision.[15]

However, as M. Ahmad notes, the main focus of the new policy was to foster loyalty to religion and to the larger Muslim *Ummah*.[16] With this overarching aim, the new policy sought to revise the existing curricula in order to bring them closer to the military regime's Islamic ideology. It also aimed to integrate *madrassah* with the secondary and post-secondary educational institutions and mobilize mosque and community based schools in order to achieve basic education and literacy goals.[17] The new policy reversed the Bhutto regimes nationalization of educational sector and textbooks were reviewed to expunge content repugnant to Islam. The military regime justified the restructuring and integrating *madrassahs* into the mainstream in the name of utilization of indigenous educational institutions. For example, the 1979 education policy stated

> In recognizing the great potential of our indigenous institutions and patronizing them for bringing about greater educational development...deviation from alien models (of education) and building upon what we already have, will make a great impact.[18]

This was clearly a politically motivated attempt on part of the military regime to institutionalize the *madrassah* and mosque schools without any attention to the structure or curricula of these institutions. The military regime particularly patronized the *madrassahs* affiliated with and operated by the religio-political parties that supported the regime. Some scholars have argued that the institutionalization of the *madrassahs* by the Zia regime was in fact a strategy of the post-colonial nationalist state to appropriate religious education and institutions with a view to transpose the post-colonial nationalism on the religious worldview of the ulema.[19] However, there are two basic flaws in this argument.

First, the military regimes such as that of General Zia-ul-Haq's do not fit neatly into the rubric of the post-colonial nationalist leadership with a modernizing mission. Military juntas are by nature insecure and illegitimate and in order to find their legitimacy they are prone to deviate from the modernizing mission of the post-colonial nationalist state. Second, it is clear that right from the beginning Zia was only interested in prolonging his rule by building a constituency for his regime. Apart from the followers of the religio-political parties that supported his regime, *madrassah* students were a resource that he was keen to tap into. The institutionalization of existing *madrassahs* and the patronage of new ones was, thus, a natural step for the military regime to take. It also suited the intelligence agencies of Pakistan and the United States as these *madrassahs* could provide a ready crop of fighters for the Afghan jihad.

The closer the *madrassahs* and their parent religio-political parties were to jihadi outfits in Afghanistan the greater importance they had for the military junta and the U.S. war effort. The radicalization of the *madrassahs* was probably inevitable. While there are no reliable figures for the growth in the number of *madrassahs* during the Zia era (1977–88) there is a general consensus that the growth was exponential. According to Zahid Hussain,

> at the height of the Afghan *jihad*—1982–1988—more than 1000 new *madrassahs* were opened in Pakistan.... [A]lmost all belonged to hardline *Sunni* religious parties like Jamiat-e-Ulema-e-Islam (JUI) and Jamaat-i-Islami (JI) which were Zia's political allies as well as partners in Afghan *jihad*.[20]

At the time of the independence of Pakistan in 1947 there were approximately 137 *madrassahs* in the country. In the next ten years the number grew to 244. The exponential increase in the number of *madrassahs* during the Zia years has continued since his death in 1988. According to a report of the Brussels based International Crisis Group (ICG), in 2002 the number of registered *madrassahs* in Pakistan was close to 10,000 with an enrollment of around 1.7 million students.[21] It is estimated that around 10–15 percent of the *madrassahs* operating in Pakistan are linked to jihadi organizations in Pakistan and abroad.

It is important to note that the *madrassahs* that have operated in India and Pakistan for hundreds of years have been a part of a very large network of informal educational institutions run for philanthropic and charitable purposes. They have also been the centers for training the clergy for different sub-sects. Their transformation into training centers for militancy and indoctrination started with the U.S. sponsored jihad in Afghanistan.[22] It is also important to note that apart from the U.S. funding and encouragement these *madrassahs* and their parent outfits also benefited from the inflow of funds from countries such as Saudi Arabia, Iraq, Iran, and Kuwait. Each supplied funds in order to maximize the influence of the sect/sub-sect it believed in and to minimize the influence of others.

While the rationale given by the Zia regime for the patronage of *madrassahs* was that it wanted to harness the network of educational institutions run on charity and philanthropic donations in order to address the problems of illiteracy in Pakistan, the fact of the matter is that the motivations of the regime in doing so were purely political. Lacking a political base of its own, the regime was keen to tap into the constituency of its allied religio-political parties. Secondly, it wanted to develop the religio-political parties as a political force to counter the political influence of the mainstream secular political parties particularly that of the Pakistan Peoples Party (PPP). Finally, the regime could draw upon the resources and contacts of the religio-political parties to fight the proxy jihad in Afghanistan. Thus, it is not surprising that little attention was given to other potential consequences of the dramatic growth of *madrassahs* and their radicalization. Similarly, the regime was simply not interested in what was being taught at these *madrassahs*. According to Tariq Rahman's impressive 2004 study of *madrassah* education in Pakistan "education in the madrassa produces religious, sectarian, sub-sectarian and anti-western bias."[23] Rahman, however, cautions that it would be too simplistic to assume that this translates automatically into militancy and violence. Other factors such as extreme poverty, the suppression of political dissent during Zia's martial law, and the military training of the young men who fought in Afghanistan and Kashmir also play an important role in this respect. Rahman's study presents worrying statistics. For example, in his wide-ranging survey of Pakistani students in public and private schools, colleges, cadet colleges, universities, and *madrassahs* on the issue of militant orientation among students and teachers, it is apparent that the students and teachers in *madrassahs* are the most militant.[24] Domestically, the students and the faculty of these *madrassahs* consider followers of other sub-sects of Islam especially the Shi'ites, upper classes as enemies of Islam and thus a jihad is warranted against them. Externally, the list of enemies includes the United States, Israel, and all governments that are allied to the United States.

While the *madrassah* factor has caught the attention of scholars and policy makers in Pakistan and abroad, a crucial development in Pakistan's educational system during the Zia years has gone relatively unnoticed. The military regime slowly and implicitly brought about changes in the public school curricula with intent to militarize the society[25] and to redefine the criteria for citizenship. I have dealt with this issue in detail elsewhere so let me briefly discuss some of the major developments in this respect.[26]

The Militarization of Curricula and Textbooks in the Zia Era

During the Zia era changes were introduced in curricula and textbooks that aimed at inculcating values related to nationalism, militarism, and war. One of the major ways in which this was done was by means of what was included in and excluded from the textbooks. At this point nationalism, militarism, and war began to be normalized by being linking to religious notions such as *jihad* (holy war) and *shahadat* (martyrdom). These textbooks are replete with stories, essays, poems, etc. about military battles from early Islamic history, wars between India and Pakistan, military heroes, and other related figures. At times 50–60 percent of the content deals with nationalist themes.[27]

There are two major types of exclusion: firstly, prominent Pakistanis other than the military heroes and the leaders of the nationalist movement are excluded from the educational discourse. There are no scientists, artists, social workers, journalists, or statesmen who are deemed worthy of inclusion in the texts. Neither the Nobel Laureate Abdus Salam, nor the acclaimed social worker Abdul Sattar Edhi (considered to be the Mother Theresa of Pakistan), is present in any textbook. Secondly, women and all religious, linguistic, or ethnic minorities are also excluded from the texts.

In addition to the first construction of the Indian/Hindu as a conniving "other" bent on undoing the existence of Pakistan, the second major construction is that the military is the most important institution in the country; military heroes are the only heroes; jihad is the most important religious duty and activity; and since all of them have a religious sanction that none of them can or should be questioned. While the constraints of space do not permit a more detailed analysis, let me briefly take some examples from the curriculum documents and textbooks produced between 1995 and 2002. For example, a curriculum document issued by the Federal Ministry of Education suggests the following for classes K–5: "A feeling be created among the students that they are members of a Muslim nation. Therefore, in accordance with the Islamic tradition, they have to be truthful, honest, patriotic and self-sacrificing warriors of Islam *(Mujahids)*."[28] Similarly, the curriculum documents mandate that the students must be made aware of the "blessings of *Jihad*" through "simple stories to incite *Jihad*"[29] and must be taught to make speeches on jihad. Other learning outcomes prescribed by the curriculum documents are to recognize the importance of jihad in every sphere of life and create love and aspiration for jihad, *Tableegh* (proselytization), *shahadat* (martyrdom), sacrifice, *ghazi* (victor of war).

While the government of Pakistan changed the curricula of public schools, a USAID-funded project at he University of Nebraska–Omaha produced special textbooks in Dari and Pashto (Dari and Pashto are major dialect of Afghanistan, while Pashto is also spoken in the bordering Pakistani province of the North-West Frontier Province [NWFP]). These textbooks aimed to inculcate values related to jihad and militant training. A very large number of such textbooks were handed out at Afghan refugee camps and *madrassahs* in Pakistani. These texts trained students in basic math by teaching them to count dead Russians and Kalashnikov rifles. Taliban later used the same books in their *madrassahs* in Afghanistan and Pakistan.[30]

To sum up, the U.S. involvement in Pakistan during the first Afghan episode brought about crucial changes in the public and informal educational sectors of the country. These changes have largely been responsible for transformations that have gnawed at the fabric of the Pakistani society and have resulted in tragic events abroad. The condoning of the mushrooming of *madrassahs* by the United States resulted in the evolution of a class of militarized young men (and women) who are anti-Western and anti-modernization. The militarization of the public school curricula has resulted in the creation of a populace that does not question the way of the military.

The United States and the Second Afghan Episode:
The War on Terror and Attempts to
De-Islamize Education in Pakistan

Just as the United States found a military dictator in Pakistan at the time of the first Afghan episode so it did after 9/11 when it wanted to wage a war on the Taliban that it helped create during the 1980s. There is however, one crucial difference. While the relationship between Pakistani army and America in 1980s centered on using "Allah" to wage a war against the Soviet troops in Afghanistan the post-September 11 relationship between the two focuses on reversing this trend. This time around de-Islamization (of education and the Pakistani populace) seems to be the mutually agreed upon goal of the Army-America nexus.

It was not surprising that for the third time in 52 years Washington has found an ally in a military general in Pakistan at a time when it needed support for a war effort.[31] What is indeed surprising is that once again Washington has failed to read through the history of domestic interludes in Pakistan, and learn from history. A reading of the democratic periods in Pakistani history, especially in the last two decades, clearly shows that it was the civilian, democratically elected governments that acted against the forces of religious obscurantism and violence, as compared to the (U.S. backed) military governments that played both sides. For example, as the ICG notes in its 2002 report, it was the democratically elected governments of Benazir Bhutto and Nawaz Sharif from 1988 to 1999 that took most concerted measures against the religio-political parties and the jihadi organizations. The logic is simple: it is in the best political interests of the mainstream political parties (that have representation in all four provinces of Pakistan) to try to minimize the political influence of religio-political parties and the jihadi organizations. The latter have never had electoral influence and the major political parties would like to keep it that way. As compared to this, the logic of military rule in Pakistan shows that due to a lack of legitimacy the army/generals always play the religio-political parties in order to gain some semblance of legitimacy. General Pervez Musharraf is no exception.

As mentioned earlier, the United States found a willing ally in general Musharaff after 9/11[32] in its desire to purge the Taliban from Afghanistan. This marriage of convenience deprived the Taliban of the support that Pakistan had been providing previously.[33] It also provided the United States with military bases, intelligence, and other support to wage the War on Terror in Afghanistan; and, it set out to deal with the now mushroomed radical *madrassah*s. In return, the Musharaff government got money and political support to prop its regime. In the following sections I will focus on the last element of the Army-America relationship in the post-September 11 era. However, one key difference between the Army-America relationship during the first afghan episode and the recent one (United States-Musharraf) must be kept in mind. The difference is that in the former relationship the United States was not overtly concerned with the Islamization of the public education curricula or the growth of the *madrassah* network. It, in fact, implicitly encouraged the Islamization of education. In the context of the recent relationship the United States has the "reform" of Pakistani educational systems as one of the key policy foci. However, it is important to note that even now the main U.S. focus in this respect is the reform of the *madrassah* system. Reforming the general Pakistani educational system seems a lesser priority for Washington.

The War on Terror and the Pakistani Educational System

As early as 2002 the Musharraf government launched education sector reforms. These reforms were focused on the public education and pledged to bring the educational spending at par with the UNESCO recommendation of 4 percent from roughly 1.7 percent,[34] raise literacy and enrollment levels, reduce dropout rates, and achieve EFA targets, and the Millennium Development Goals.

Soon afterward the United States started to channel money for the reform of primary education and literacy in Pakistan. In 2003–4 USAID pumped in $28 million and another $66 million in 2005. For 2008 the State Department Congressional Budget has requested $52 million for general educational programs, an additional $50 million for the earthquake reconstruction, and $110 million for development of Federally Administered Tribal Areas.[35] This aid, though welcome and needed, seems pittance in relation to approximately $10 billion in military aid and fund to fight War on Terror that United States has given to the government of Pervez Musharraf.

As mentioned earlier, the bulk of U.S. aid for educational reform has been diverted to fighting/reforming/controlling the *madrassahs*. However, the Musharaff government has also tried, albeit unsuccessfully, to reform the public school curricula and to expunge militaristic content. These efforts however, have fallen victim to political expediency. For example, the government backtracked on its effort to rid the textbooks of the militant content when a political storm brewed over the expunging of Hadith related to jihad or Shahadat from eleventh and twelfth grade biology textbooks. Muttahida Majlis-e-Amal (MMA), a conglomeration of religio-political parties in the National

Assembly, walked out of the National Assembly. Liaquat Baloch, of the Jamaat-e-Islami party alleged that it was "under the conditionalities of the U.S. Agency for International Development, [that] all verses containing any reference to jihad or exposing anti-Muslim prejudices of Jews and Christians are being omitted from the syllabi."[36] The Jamaat-e-Islami chief threatened to bring a privilege motion[37] against the Pakistan Muslim League (Q) (PML-Q) government that had been propped up by the military regime.

The Musharraf government, contrary to its position that it would rid education and the society of the militant religious elements, thought it expedient to go on a defensive and present an apologetic position. For example, Zubaida Jalal, the Federal Education Minister claimed that the Hadith and verses on Shahadat and jihad had merely been shifted from eleventh and twelfth grade biology textbooks to tenth grade biology textbooks.

Similarly, when the report by the Islamabad-based Sustainable Development Policy Institute on curricula in Pakistan came under fire from the religio-political parties, especially the Jamaat-e-Islami, the government and the education minister Zubaida Jalal withdrew their blessings from the report and claimed that the report has been rejected by the 15 member government review committee when it had in fact approved the report by a vote of 9–6.[38]

In a similar vein, the government shelved a report titled "The National Curriculum 2000: A Conceptual Framework" that it had commissioned earlier. In private conversations that I had with people at the Ministry of Education and European Survey Research Association it was reported that neither the government nor the USAID was willing to touch the curriculum issue in either the realm of public education or of the *madrassah*.

Madrassah Reform

Both the United States and the military government of Pervez Musharraf concurred that *madrassahs* in Pakistan were the first and foremost source of terrorism and militancy in Pakistan and beyond, and that these had to be tamed, reformed, or simply uprooted and banned if need be. However, it should be noted that according to most scholars and sources it is only a tiny minority of around 10–15 percent of the *madrassah* that are radicalized and actively engaged in producing a jihadi mindset among its students. A large majority of *madrassahs* is still carrying on the centuries old tradition of housing, feeding, and educating the children of the destitute of the Pakistani society. As noted above, according to the Brussels based ICG there are over 10,000 *madrassahs* operating in Pakistan. Some scholars place the number at 13,000.[39] There is a disagreement as to the exact number of students enrolled in these *madrassahs*. The World Bank puts the number at 475,000[40] while according to the ICG the number of students enrolled in *madrassahs* is between 1.5 and 1.7 million.[41]

The effort to counter the monster of religious extremism and militarism that operates out of the radicalized *madrassahs* has not been very successful. The United States and the world community largely blame the Musharraf government of evading the issue and not acting strongly and meaningfully to contain this menace. The ICG in its 2002 report wrote,

> Musharraf's government has...relied mostly on cosmetic measures to advance its stated goal to crackdown on militants and reform madrasas. Since international pressure rather than a desire for change has shifted its stance, the government remains reluctant to initiate fundamental changes in the very policies it promoted that have spawned militancy.[42]

A recent CNN documentary: "Pakistan: The Terror Central," reached similar conclusions.[43] General Pervez Musharraf on the other hand has consistently maintained that he has made concerted efforts in this respect by banning jihadi organizations, freezing their bank accounts, instituting tighter control over money transmissions from abroad, keeping stricter watch over foreigners in the *madrassahs*, opening three model *madrassahs*, establishing a *madrassah* registration law, and providing financial incentives to *madrassahs* that did register. While the blame-game goes on, it will not be too far-fetched to say that there is a gap between the desire and effort on both sides (United States as well as the Musharraf government).

Two crucial elements are needed in the attempt to harness the radical *madrassahs*. First, there has to be a strong political will. And secondly, the resources needed must be available. It is evident that for General Musharraf the political will is subservient to his desire to hang on to power. Thus, the efforts to control the jihadi *madrassahs* are at best halfhearted and the results mixed. On the other hand, it seems that the United States wants a quick fix, an ad hoc solution to a very complex problem. It seems to favor a carrot and stick approach, that is, offering incentives to the *madrassahs* or clamping down if they don't accept. The incentives, as mentioned earlier, are paltry in comparison to the overall War on Terror budget. For example, in 2001–2 approximately 16 million rupees (approximately US$250,000) were earmarked for *madrassahs* that accepted the government incentive. This according to Tariq Rahman amounted to Rs. 1.55 per student/year. For 2003–4 the outlay was increased to Rs. 30.45 million or Rs. 28.60 (around 50 cents US) per student/year.

According to a 2002 report of the ICG, the Ministry of Education sought US$233 million as an incentive package to lure *madrassahs* to register under the new law to modernize their curricula. The package had several components: a one time grant of $20,000/million over three years to cover free books and ten school cabinets; funds to cover the cost of hiring teachers for formal subjects; computers (five for each *madrassah* and one printer); and teacher training envisioned as 32,000 teachers being hired for three years. According to ICG 2002 "the entire project is either bureaucratic wishful thinking or an exercise in PR for a Western audience."[44] In the light of the above, it is not difficult to see that neither the military government in Pakistan nor the United States seems to be serious in a long-term and durable solution to the menace of militarist *madrassahs*.

Conclusion

The U.S. relationship with Pakistan at various points in the history has been directed by short term geostrategic interests. It has never developed or evolved into a mutually beneficial relationship over the long term. While this relationship has been a focus of scholarly research, the impact of this ad hoc, on-again, off-again relationship on Pakistan's educational system has never been a focus of serious research. It is only now that researchers as well as the U.S. policy makers are paying attention to this important aspect. The reasons for this attention lie in the context of the U.S.-led War on Terror against a foe whose very existence is in large part due to changes in Pakistan's educational system that the United States encouraged both directly and indirectly in the 1980s. However, even now, as mentioned above, this attention is piecemeal, ad hoc and myopic. With its focus on reigning in the radicalized *madrassahs*, the policy makers in both Washington and Islamabad seem to be suffering from tunnel vision.

If the menace of the radicalized *madrassahs* and their product—the jihadi mindset—has to be arrested and pushed back it would require a concerted effort based on a broader and holistic vision of the problem.

The problem is not only the *madrassahs*; had that been the case it would have been relatively easier to harness it through policing and military means. The real problem is that the military regimes in cohort with their allies in Washington encouraged the constitution of militarized subjects and subjectivities through militarization of curricula and textbooks in Pakistan's general public education system. These subjects, as mentioned earlier, are taught to extol the military values. Thus it is not surprising that they tolerate the existence of the *madrassahs* and the jihadi mindset created by them.

In my view, since the United States was actively involved in the creation of this monster, it must now actively contribute to its dismantling. The United States must encourage and proactively help Pakistan in reforming the educational sector so that it rids itself of the militarist discourses and instead gears up to constituting democratic citizens. The first step in this respect will be for the Untied States to withdraw its support from the military regimes even if they are under the garb of semi-civilian set ups. It is clear that historically it has been the democratically elected regimes that have displayed a greater potential and zeal for dealing with the jihadis than the military regime of General Pervez Musharraf. Democratically elected civilian regimes once allowed to operate without the fear of military intervention will be in a much better position to deal with the militant *madrassahs*.

Finally, the *madrassah* reform has to be clearly thought out. Questions such as whether we want *madrassahs* to become a part of the public education system must be raised and deliberated upon. An affirmative answer to this question, for instance would mean that additional resources would have to be allocated to a sector that is already struggling because of inadequate budgetary allocations. Similarly, if the *madrassahs* remain outside the purview of state's oversight then it is doubtful that they can be encouraged to teach not only the religious subjects but also the secular curricula. These questions (among other) must be deliberated and debated upon at the societal level in Pakistan in order to re-articulate the relationship between Allah, America, and the army in Pakistan.

A Pakistan where the army defends the borders (and not subvert the democratic process), where Allah provides, and organizing framework for day-to-day life (and not only the zeal for jihad) will be a Pakistan that would bring stability to the region. Education reform is the key to this reorientation of the 3-A trinity.

Notes

1. *Madrassah*, an Urdu word literally means a school or a place of learning. In South Asia, (especially in India, Pakistan, and Bangladesh) it is used to refer to religious schools usually associated with a sect or sub-sect of Islam. In Urdu, the plural of *Madrassah* is *Madaris*. However, in this paper I use the more common anglicized usage of adding an s after *madrassah* to denote the plural form.

2. For details on these treaties and alliances see M. Ayaz Naseem, *Pak-Soviet Relations: 1947–1965* (Lahore, Pakistan: Progressive Publishers, 1988).

3. Despite being a valued ally Pakistan did not get the support from the United States that it had expected. The U.S. promise of dispatching the sixth fleet to ward off Indian attacks from the sea never materialized.

4. *Mujahideen* is the plural of *mujahid*, which means "a warrior of the God."

5. See Pervez Musharraf, *In the Line of Fire: A Memoir* (London, New York: Simon and Schuster, 2006).

6. Ministry of Education website, 2004. www.moe.gov.pk.

7. I have dealt with this in detail elsewhere. Please see M. Ayaz Naseem and Adeela Arshad-Ayaz, "The Market, the Nation and the School: EFA in Times of Globalization and Nationalism," in *Education for All: Global Promises, National Challenges*, ed. Alexander Wiseman and David Baker (Amsterdam: Elsevier Science, Ltd., 2007), 73–108.

8. Concurrent subjects are those that are under the purview of both the federal and the provincial governments.

9. Fred Halliday, *The Making of the Second Cold War* (London: Verso, 1983).

10. For the so-called Islamization of law see Asma Jahangir and Hina Jillani, *Hudood Ordinances: A Divine Sanction?* (Lahore, Pakistan: Rohtas Books, 2003). Shahnaz Rousse, "Gender(ed) Struggles: The State, Religion and Society," in *Against All Odds: Essays on Women, Religion and Development from India and Pakistan*, ed. Kamla Bhasin, Ritu Menon, and Nighat Said Khan (New Delhi: Kali for Women, 1994). Various articles In Hassan Gardezi and Jamil Rashid eds., *Pakistan, the Roots of Dictatorship: The Political Economy of a Praetorian State* (London: Zed Press, 1983). Nighat Said Khan and Afiya Sherbhano Zia, *Unveiling the Issues: Pakistani Women's Perspectives on Social, Political and Ideological Issues*, transcribed and trans. Naureen Amjad and Rubina Saigol (Lahore, Pakistan: ASR Publications, 1994). For the Islamization of the economy see Mohammad Waseem, *Politics and the State in Pakistan* (Islamabad: National Institute of Historical and Cultural Research, 1994). Also see Oman Noman, *Pakistan: A Political and Economic History Since 1947* (London: Kegan Paul International, 1990).

11. See M. Ayaz Naseem, "The Soldier and the Seductress: A Poststructuralist Analysis of Gendered Citizenship through Inclusion in and Exclusion from Language and Social Studies Textbooks in Pakistan," *International Journal of Inclusive Education* 10, nos. 4–5 (2006): 449–468. Also see M. Ayaz Naseem, "Deconstructing Militaristic Identities in Language and Social Studies Textbooks in South Asia: The Case of Pakistan," in *Caught in the Web or Lost in the Textbook?* ed. Éric Bruillard, Bente Aamotsbakken, Susanne V. Knudsen and Mike Horsley (Paris: Jouve, STEF, IARTEM, IUFM de Basse-Normandie, 2006), also available at: http://caen.iufm.fr/colloque_iartem/pdf/part_I.pdf (accessed January 10, 2009).

12. Zahid Hussain, *Frontline Pakistan: The Struggle with Militant Islam* (New Delhi: Penguin/Viking, 2007).

13. The Reagan administration consistently condoned these discriminatory infarctions. See for example Mohammad Waseem, *Politics and the State in Pakistan*, 2nd ed., 383–385.

14. M. Ahmad, "Education" in *Pakistan in Perspective 1947–1997*, ed. Rafi Raza (Karachi: Oxford University Press, 2001), 247.

226   *M. Ayaz Naseem*

15. Government of Pakistan, Ministry of Education, *New Education Policy 1978* (Islamabad, 1978).
16. Ahmad, "Education."
17. Ibid., 247.
18. Government of Pakistan, Ministry of Education, National Education Policy and Implementation Programme. Islamabad: Printing Corporation Of Pakistan, 1979.
19. Jamal Malik, *Colonization of Islam: Dissolution of Traditional Institutions in Pakistan* (Lahore, Pakistan: Vanguard, 1996). *Ulema* is the plural of *Alim*, which means (Islamic) scholar.
20. Hussain, *Frontline Pakistan*, 80.
21. ICG, "Pakistan: Madrassahs, Extremism and the Military," *ICG Asia Report*, no. 36 (Islamabad/Brussels: ICG, 2002), 2.
22. Hussain, *Frontline Pakistan*, 77.
23. Tariq Rahman, *Denizens of Alien Worlds: A Study of Education, Inequality and Polarization in Pakistan* (Karachi: Oxford University Press, 2004), 89.
24. For instance, responding to the questions on whether Pakistan should "take Kashmir away from India by an open war or by supporting Jihadi groups or by peaceful means?" 59.86 percent of the *madrassah* students responded in favor of war, 52.82 percent in favor of supporting jihadi groups. Among *madrassah* faculty, 70.37 percent were in favor of open war and 59.26 supported jihadi groups as second option. Ibid., 174–176.
25. By militarization, I mean constituting subjects and subjectivities that do not question the ways of the military.
26. See Naseem, "The Soldier and the Seductress," 449–468.
27. Khursheed Kamal Aziz, *The Murder of History: A Critique of History Textbooks Used in Pakistan* (Lahore, Pakistan: Vanguard, 1993).
28. Government of Pakistan, Ministry of Education, *Curriculum Document Primary Education, Classes K-V* (Islamabad: Curriculum Wing, 1995).
29. Ibid.
30. Hussain, *Frontline Pakistan*, 80.
31. General Ayub Khan in 1950s at the time of the Korean War in particular and the U.S. policy of containment in general; General M. Zia-ul-Haq in 1980s at the time of the first Afghan episode and now General Pervez Musharraf at the time of the U.S. War on Terror.
32. Musharraf, *In the Line of Fire*, 201. In his memoir General Musharraf notes that he had little choice after the September 11 attacks but to stop supporting the Taliban and ally with the U.S.-led War on Terror. According to Musharraf, "In what has to be the most undiplomatic statement ever made, Armitage…told the director general not only that we had to decide whether we were with America or with the terrorists, but that if we chose the terrorists, then we should be prepared to be bombed back to the Stone Age." Armitage denies ever having made the statement.
33. For details of Pakistan's support to Taliban see Ahmed Rashid, *Taliban: Militant Islam, Oil and Fundamentalism in Central Asia* (New Haven, CT: Yale University Press, 2000).
34. Spending on education has been on a steady decline from 3.2 percent in 1990s to 1.7 percent currently.
35. Lisa Curtis, "US Aid to Pakistan: Counting Extremism Through Education Reform," *Heritage Lectures* 1029 (2007): 2.
36. Massoud, Ansari, "Lessons in Intolerance," *Newsline*, May 2004, http://newsline.com.pk/NewsMay2004/NewsspMay1.htm (accessed January 10, 2009).
37. Any member from either Legislative Assembly can move a Privilege Motion notice against a person or the government who he/she feels has caused breach of privilege either of his own person or that of the house in general.
38. Ibid.
39. Hussain, *Frontline Pakistan*, 79.

40. Tahir Andrabi, Jishnu Das, Asim Khawaja, and Tristan Zajonc, "Religious School Enrollment in Pakistan: A look at the Data," Working Paper Series 3521, World Bank, February 1, 2005.
41. ICG, "Pakistan: Madrassahs, Extremism and the Military," 2.
42. Ibid., 19.
43. CNN Live Special, "Special Investigations Unit—Pakistan Terror Central," Aired December 27, 2007—23:00 ET. Transcript of the documentary is available at: http:// transcripts.cnn.com/TRANSCRIPTS/0712/27/se.01.html (accessed January 10, 2009).
44. ICG, "Pakistan: Madrassahs, Extremism and the Military," 26.

Bibliography

Ahmad, Muneer, "Education" in *Pakistan in Perspective 1947–1997*, ed. Rafi Raza, 238–275. Karachi: Oxford University Press, 2001.

Aziz, Khursheed Kamal. *The Murder of History: A Critique of History Textbooks Used in Pakistan.* Lahore, Pakistan: Vanguard, 1993.

Curtis, Lisa. "U.S. Aid to Pakistan: Counting Extremism through Education Reform." *Heritage Lectures,* no. 1029 (2007), available at: http://heritage.org/research/asiaandthepacific/hil1029.cfm (accessed January 15, 2009).

Gardezi, Hassan and Jamil Rashid, eds. *Pakistan, the Roots of Dictatorship: The Political Economy of a Praetorian State.* London: Zed Press, 1983.

Halliday, Fred. *The Making of the Second Cold War.* London: Verso, 1983.

Hussain, Zahid. *Frontline Pakistan: The Struggle with Militant Islam.* New Delhi: Penguin/Viking, 2007.

Khan, Nighat Said and Afiya Sherbhano Zia. *Unveiling the Issues: Pakistani Women's Perspectives on Social, Political and Ideological Issues.* Transcribed and Trans. Naureen Amjad and Rubina Saigol. Lahore, Pakistan: ASR Publications, 1995.

Malik, Jamal. *Colonization of Islam: Dissolution of Traditional Institutions in Pakistan.* Lahore, Pakistan: Vanguard, 1996.

Musharraf, Pervez. *In the Line of Fire: A Memoir.* London: Simon and Schuster, 2006.

Naseem, M. Ayaz. *Pak-Soviet relations: 1947–1965.* Lahore, Pakistan: Progressive Publishers, 1988.

———. "The Soldier and the Seductress: A Post-structuralist Analysis of Gendered Citizenship through Inclusion in and Exclusion from Language and Social Studies Textbooks in Pakistan." *International Journal of Inclusive Education* 10, nos. 4–5 (2006): 449–468.

———. "Deconstructing Militaristic Identities in Language and Social Studies Textbooks in South Asia: The Case of Pakistan," in *Caught in the Web or Lost in the Textbook?* Ed. Éric Bruillard, Bente Aamotsbakken, Susanne V. Knudsen and Mike Horsley. Paris: Jouve, STEF, IARTEM, IUFM de Basse-Normandie, 2006.

Naseem, M. Ayaz and Adeela Arshad-Ayaz. "The Market, the Nation and the School: EFA in times of Globalization and Nationalism," in *Education for All: Global Promises, National Challenges,* ed. Alexander Wiseman and David Baker, 73–108. Amsterdam: Elsevier Science, Ltd., 2007.

Noman, Omar. *Pakistan: A Political and Economic History Since 1947.* London: Kegan Paul International, 1990.

Rahman Tariq. *Denizens of Alien Worlds: A Study of Education, Inequality and Polarization in Pakistan.* Karachi: Oxford University Press, 2004.

Rashid, Ahmed. *Taliban: Militant Islam, Oil and Fundamentalism in Central Asia.* New Haven, CT: Yale University Press, March 2000.

Rousse, Shahnaz. "Gender(ed) Struggles: The State, Religion and Society," in *Against all odds: Essays on Women, Religion and Development from India and Pakistan*, ed. Kamla Bhasin, Ritu Menon, and Nighat Said Khan, 16–34. New Delhi: Kali for Women, 1994.

Waseem, Mohammad. *Politics and the State in Pakistan*. 2nd ed. Islamabad: National Institute of Historical and Cultural Research, 1994.

CHAPTER ELEVEN

Corporate Education and "Democracy Promotion" Overseas: The Case of Creative Associates International in Iraq, 2003–4

KENNETH J. SALTMAN

According to neoconservative scholars as well as their critics, the events of September 11, 2001 allowed the implementation of premade plans to radically reshape the U.S. national security strategy to pursue more aggressively and openly global military and economic dominance and to force any and all nations to submit to a singular set of American values.[1] With the declaration of military response, the United States invaded first Afghanistan (2001) and then Iraq (2003), in part, on the justification that these were fronts in the so-called War on Terrorism.[2] Following both invasions the United States, through the Agency for International Development (USAID), contracted with a private for-profit corporation, Creative Associates International, Inc. (CAII) to lead the rebuilding of education. School buildings, textbooks, teacher preparation, curriculum planning, administration—all would be implemented by CAII directly or by firms subcontracted by CAII. In 2003, the company came under close scrutiny by Congress and the press for receiving its Iraq contract without competitively bidding for it.[3] The no-bid contract with Creative Associates International was one of a number of no-bid contracts benefiting U.S. corporations including Bechtel (which has been subcontracted by CAII to build schools), Haliburton, and others that profited from rebuilding in Iraq.[4]

What is at stake in the case of CAII in Iraq is not principally a matter of proper bidding protocol in educational contracting, nor even merely the possibility that CAII was involved in "war profiteering,"[5] that the rebuilding "looks like a criminal racket,"[6] and that it is reaping "the windfalls of war."[7] The role of CAII in remaking education in Central Asia, the Middle East, and around the world on behalf of the United States concerns a number of broader issues. In one sense CAII represents just one kind of international corporate involvement in schooling: educational development. Yet, the array of for-profit projects that CAII is involved with in Iraq, Afghanistan, and around the globe makes it exemplary of a range of global corporate schooling initiatives including textbook production, curriculum design, remediation services, teacher education programs, and privatization schemes. CAII has been involved since the beginning of the "Reagan revolution" in "democracy promotion" projects that merged development work with political, military, and economic influence strategies on the part of the United States. Such work included reintegrating Contra terrorists into Nicaraguan civil society through work training; influencing Nicaraguan elections; participating in both coups against Aristide in Haiti; and privatizing, commercializing, and Americanizing Haitian media and journalism particularly around election coverage. I am concerned here with the changing relationship between nation-states, corporations, and education, as the United States under the Bush administration continues its neoconservative foreign policy that emphasizes the use of military force to install what its advocates describe as capitalist democracies modeled on the U.S. system.[8] Corporate education appears to have a central role in the neoconservative model as CAII appears on stage to rebuild following these military actions.[9]

Though there is much new about the present political constellation, CAII's history, for example, in support of the Contra guerillas in Nicaragua, highlights continuities in the role of education in aggressive U.S. foreign policy interventions that are favorable to U.S.-based transnational capital. As the case of CAII illustrates, corporate educational development experts appear integral to U.S. economic and military strategy around the world. As the United States was developing a more sophisticated strategy to influence political process and educational apparatuses in the 1980s, CAII was there and has continued to be there funded by USAID and working in conjunction with other corporations and nonprofit organizations.[10] In addition, we can note that CAII's 2003 takeover of educational development in Afghanistan signified a break with the longstanding role that the public and nonprofit

University of Nebraska had played since the 1970s in fundamentalist Islamic schools—work that, as Naseem notes in this volume, helped to create the Mujahedeen that served the U.S.'s "market fundamentalist" project of driving out the Soviets. These Mujahedeen "good pupils" of Nebraskan texts would then become the nemesis of the United States in the form of the Taliban and Al-Quaeda.

CAII's recent role in Iraq also signifies something new. If, the Iraq War was in part a radical free market experiment bent on demolishing the public sector and shifting control of civil society nearly completely to the private sector, then education was not only a political and ideological concern of the United States as in Afghanistan in the 1970s and 1980s. As Naomi Klein, Christian Parenti, Pratap Chatterjee, and others have argued, the Iraq War has been a radical experiment in widescale neoliberal privatization—an attempt to essentially hand a nation over to corporations.[11] They have also suggested that the military resistance to the United States in Iraq has been inextricably tied to attempts to retain control over industry and labor. Within this view, education is, on the one hand, just another business opportunity provided for by war. And, on the other hand, it is an experiment with the conservative U.S. domestic policy agenda of educational privatization that includes vouchers, charter schools, performance contracting, for-profit remediation, as well as the broad spectrum of educational reforms that are designed to set the stage for these privatization initiatives including performance-based assessment, standardization of curriculum, and recourse to so-called scientific-based educational research.[12] But CAII, USAID, and the Department of Defense do not openly admit that their projects are foremost a matter of promoting a U.S. brand of capitalism. Rather, these projects are defined as a form of "democracy promotion."

Such "democracy promotion" projects contain elements of neoliberal ideology in that they conflate economic values and political values while they ultimately exist to promote forms of political governance and modes of political subjectivity conducive to neoliberal economic policy. However, the neoconservative uses of "democracy promotion" projects differ from a Clinton-era type of neoliberal thought principally by emphasizing the use of coercion in the form of military power where the "Clintonians" emphasized economic influence foremost.[13] Neoconservatism continues and intensifies the neoliberal model of diminishing the caregiving roles of the state while strengthening the repressive and punitive roles. This recourse to nationalism defines nationalism through consumerism, individualism, jingoistic patriotism, and with neoliberal terms such as Bush's "ownership society" that retains the ideal of the entrepreneurial subject, and that rejects values of social responsibility and civic participation. Yet, overseas "democracy promotion" projects have since their inception employed distinctly public-minded terms such as "civic education," "civic participation," "electoral reform," and "democratic media reform."

As William I. Robinson has argued, "In elaborating a policy of 'democracy promotion,' the United States is not acting on behalf of a 'U.S.' elite, but playing a leadership role on behalf of an emergent transnational elite."[14] This model of capitalist democracy requires political and economic reform consistent with the interests of capital in the richer nations. This includes privatization of state-controlled industry, deregulation of rules protecting domestic markets and a number of civil society reforms consistent with these changes: (1) a depoliticized populace, (2) individualized consumer-oriented subjects who are decreasingly inclined to identify their interests with collective social action and increasingly inclined to identify their interests with consumption practices,[15] (3) a political subject friendly to heavy foreign involvement by powerful states and organizations, (4) reliance on expensive foreign-provided technology in the place of broad-based participatory civic involvement, (5) the reform of media and journalism on the private U.S. model, (6) the reform of political elections as intertwined with for-profit media on the U.S. model, and (7) an emphasis on privatization of state-run knowledge-making institutions.

In a theoretical sense there is a question as to why the Bush administration, so bent on the pursuit of power through coercion, would also be so focused on producing hegemonic consent through the use of "democracy promotion" projects within nations targeted for military attack. Yet, concern for education, public opinion, and knowledge-producing institutions like schools and mass media appear to remain of high concern to an administration that has nonetheless shifted to a more overt use of force to achieve policy aims. On one level the explanation is that the Bush administration and the neoconservatives generally (quite unlike the Clintonians) understand the centrality of pedagogy to politics. They want to hand the environment over to polluting corporations or privatize social security or invade a nation and they attempt to educate the public as to the virtue and necessity of doing so. This approach to governance is distinct from the longstanding approach of the Democratic Party to build policy, based on marketing feedback from voters rather than moral vision and political ideal. Of course, one can suggest that the moral visions and political ideals of the neoconservatives are more retrograde jingoism than principle and more often than not are pretty propagandistic wrapping for corporate plunder. Nonetheless, the neocons selectively grasp the Gramscian insight that hegemony requires leadership and political leadership demands educating those who are being led. Of course, the neocons have cared very little about making consent with the outer shell of foreign policy. As Giovanni Arrighi has written, hegemony is unraveling in the sense that the United States has lost its ability to lead other nations and is left with coercion. And yet when the United States uses coercion, as in Iraq (invasion) or Haiti (coup), it follows with "democracy promotion."

To understand "democracy promotion" in relation to coercion and consent at a deeper level it is instructive to employ the still relevant distinction made by Louis Althusser who was influenced by Antonio Gramsci—both of whom were influenced by and wrote about Machiavelli, a major figure of influence on the neoconservatives. Althusser viewed the state as an arm of capital, yet emphasized Gramsci's recognition that the struggle for leadership of civil society is a crucial strategic, political, and pedagogical one leading up to the revolutionary seizure of the bourgeois state (Ideological State Apparatus are both "stake and site" of struggle). Althusser distinguished between the Repressive State Apparatus (RSA) (military, police, judicial system) and the Ideological State Apparatus (ISA) (schools, media, religions) to illustrate how the ruling capitalist class maintains control of the reproduction of the conditions of production through coercion and consent. He also emphasized that the RSA has crucial ideological components (the culture of the military matters decisively for repressive power to hold) while the ISAs contain crucial repressive elements (the disciplinary culture of the school keeps kids there to learn know-how in forms that are ideologically consistent with the ruling interests). Although there are numerous problems with many aspects of Althusser's thought from the "scientism" to the class reductionism to, the many other problems accompanying the Marxist legacy, to the failure to theorize sufficiently a theory of agency, nonetheless he offers important tools for understanding the wielding of state power and the ways it has been increasingly used to undermine public democratic power in the United States and around the world.[16] One crucial question that he offers us now is what accounts for the shift in the wielding of state power from ISA to RSA—that is, from consent to coercion. To put it simply, why has there been a shift toward increasing use of overt repressive force on the part of the United States in foreign and domestic policy following an era of neoliberalism that emphasized the enforcement of the Washington Consensus principally through economic sanction and threats of economic and military force?[17] To complicate matters, why has U.S. civil society seen a rising culture of militarism that extends through popular culture and mass media and education? Perhaps most crucially, what should we make of the concommittent rise in ISA activity internal to states targeted for "The New Imperialism"? In education the United States is militarizing its own schools in numerous ways[18] while transforming Iraqi schools, at least in rhetoric, on the model of liberal and progressive education. As Althusser explained, RSAs have constitutive ideological components (the culture of the military) and ISAs have constitutive repressive dimensions (the compulsory attendance of schooling) yet why recent shifts? David Harvey offers a compelling economic argument for the general shift to repression explaining the shift from neoliberalism to neoconservatism: that neoliberal policy was coming into dire crisis already in the late 1990s as deregulation of capital was resulting in a threat to the United States as it lost the manufacturing base and increasingly lost service sector and financial industry to Asia. For Harvey the new militarism in foreign policy is partly about a desperate attempt to seize control of the world's oil spigot as lone superpower parity is threatened by the rise of a fast growing Asia and a unified Europe with a strong currency. Threats to the U.S. economy are posed by not only the potential loss of control over the fuel for the U.S. economy and military but also the power conferred by the dollar remaining the world currency, the increasing indebtedness of the United States to China and Japan as they prop up the value of the dollar for the continued export of consumer goods. For Harvey, the structural problems behind global capitalism remain the Marxian crisis of overproduction driving down prices and wages while glutting the market and threatening profits and the financialization of the global economy. Capitalists and states representing capitalist interests respond to these crises through Harvey's version of what Marx called *primitive accumulation,* "accumulation by dispossession." As Harvey explains privatization is one of the most powerful tools of accumulation by dispossession, transforming publicly owned and controlled goods and services into private and restricted ones—the continuation of "enclosing the commons" begun in Tudor England.

There is a crucial tension presently between two fundamental functions of public education for the capitalist state. The first involves reproducing the conditions of production: teaching skills and know how in ways that are ideologically compatible with the social relations of capital accumulation. Public education, whether in the United States or Iraq, in this sense remains an important and necessary tool for making political and economic leaders or docile workers and marginalized citizens or even participating in sorting and sifting out those to be excluded from economy and politics completely. The second function that appears to be relatively new and growing involves the capitalist possibilities of pillaging public education for profit whether in the United States, Iraq, or elsewhere. Drawing on Harvey's explanation of accumulation by dispossession we see that in the United States the numerous strategies for privatizing public education from voucher schemes, to for profit charter schools, to forced for-profit remediation schemes, to dissolving public schools in poor communities and replacing them with a mix of private, charter, and experimental schools follows a pattern of destroying and commodifying schools where the students are redundant to reproduction processes while maintaining public investment in the schools that have the largest reproductive role of turning out managers and leaders. These strategies of capitalist accumulation, dispossession and reproduction, appear to be at odds yet they feed each other in several ways: exacerbating differentiation and hierarchization in an ideological apparatus such as education or media through privatization and decentralization weakens universal provision, weakens the public role of a service, puts in place reliance upon expensive equipment supplied from outside, and then justifies further privatization and decentralization to remedy the deepened differentiation and hierarchization that has been introduced or worsened through privatization and hierarchization. The obvious U.S. example is the failure of the state to properly fund public schools in poor communities and then privatize those schools by turning them over to be run by corporations.[19] Rather than addressing the funding inequalities and the intertwined dynamics at work in making poor schools the remedy is commodification. Such a "smash and grab" approach to ideological state apparatuses appeared, as we will see, in Iraq as infrastructure was devastated through sanction and war and followed up with privatization and decentralization. A pattern of disaster-based public education privatization appears not only in Iraq but also in U.S. cities, such as in New Orleans where the disaster of hurricane Katrina was used to implement the largest ever voucher scheme. It is also apparent in the razing of public housing across U.S. cities to be replaced by smaller scale public private partnership "mixed income" developments in conjunction with the dismantling of public schools in those communities. These new schools and their higher test scores are being hailed as proof of the virtues of experimentation and deregulation when are in fact they are part of gentrification and the displacement of poor and minority students from coveted real estate enclaves with richer and whiter students with higher cultural capital.

The operations of Creative Associates International raise a number of crucial questions about the role of the corporation in U.S. foreign policy: How are democratic commitments being defined and enacted in foreign education and media aid? To what extent is educational development an investment in potential inexpensive labor force for U.S. based capital and the formation of a consumer base for U.S. corporations under the guise of national security? To what extent is the project of corporate globalization being implemented through military action while being redefined through the discourses of personal safety and democratic ideals? The over-arching and related question I raise through the example of CAII is whether such "democracy promotion" projects are best understood as fostering or hindering the expansion of democratic social relations in the areas of politics, economy, and culture. I suggest that progressive educators ought to link global corporate schooling initiatives with broad geopolitical questions, cultural politics, and pedagogical approaches that offer new modes of interpretation that can become acts of intervention.

Background: Creative Associates International, Inc.

Creative Associates Inc. was founded in 1979 by four women who had been partners in a day-care business. Maria Charito Kruvant, Ilda Cheryl Jones, Diane Trister Dodge, and Mimi Tse began the company in Kruvant's basement as a for-profit management consulting company through the Small Business Administration's minority-owned business program. Creative Associates got its first contract with USAID to "help poor women" in Kruvant's native Bolivia and brought in less than $100,000 in its first year.[20] When in 1983 Jones left the firm the partners changed the name to Creative Associates International, Inc. By 1985 CAII was a multimillion dollar business with government and business contracts. It has received more than 400 contracts around the world with offices in 11 countries, more than 300 employees, and annual revenue as high as $50 million.[21] It works or has worked in Angola, El Salvador, Haiti, Afghanistan, Jordan, Benin, Guatemala, Lebanon, Liberia, Mozambique, Nicaragua, South Africa, Peru, Serbia, and Montenegro. Ninety percent of its revenue comes from USAID while clients include the U.S. Marine Corps and the World Bank. According to the *Washington Times*, Kruvant and her cofounders started the company to move from the nonprofit sector into the business world and make "at least a little money doing development work."[22] Kruvant and CAII have made quite a bit more than "a little money" from CAII's contracts in Iraq alone making over $100 million and an additional $83 million if extended beyond 2 years. Kruvant has come to own 69 percent of CAII with Mimi Tse owning the other 31 percent.

Born in La Paz, Kruvant is the daughter of a wealthy landowning family forced to flee to Argentina in 1955. In 1963 Kruvant studied in New Jersey, went to Argentina to earn a teaching degree and then returned to New Jersey to work with "disadvantaged children."[23] "She was involved in passing federal legislation to promote bilingual education, founded centers for bilingual education in several states and helped develop bilingual education programs in the Washington, DC area and New York."[24] Kruvant's early inclination for liberal if not progressive educational perspectives appears on the surface to extend to CAII's contemporary work. Robert Gordon, CAII's director of operations, described the company's work in Iraq as involving not just assessing what needs to be done with the education system, rebuilding schools, redesigning curriculum, and developing teacher training. But also, as Gordon stated, "We want them to get away from rote learning. We want students to be able to ask questions."[25]

The popular press has seized on CAII's educational development work in Iraq and the fact that it is a minority-owned business as the ultimate argument against progressive and left-wing criticisms of the Iraq War as an imperial oil war waged for the strategic and economic benefit of the United States.[26] *The Economist*, which admits that the war profiteering is much broader than oil, nonetheless relies on CAII to suggest that the profiteering is hardly driven by cronyism. In a section called "Phony Cronyism" the magazine wrote a month after the U.S. invasion,

> In truth, the bidders are a broad church...Charito Kruvant, president of Creative Associates International, a "minority, women-owned and managed firm," based in Washington, DC, that is said to be USAID's preferred choice to revamp Iraq's education system does not sound like a typical conservative crony of Mr. Cheney. She signs the firm's "message from our president" with "Peace, Charito."[27]

If CAII is used in the popular press as evidence that the war was not driven by crony capitalism, it is also used to highlight the ethnic diversity and inclusiveness of the corporate economy. The magazine *Hispanic Business* acclaims CAII as a minority success story ranking it number 113 on the "Hispanic Business 500."

Idealism, peace, critical thinking, multiculturalism—what then is CAII? A group of well-meaning progressive educators who have discovered how to have their efforts in war-torn regions be well-remunerated? Are Kruvant, Tse, Gordon, Horblitt, and company a cohort of idealists merely responding to political events to further their goals and ideals under the umbrage of the U.S. pursuit of its strategic interests and national security strategy? Or rather is CAII and global U.S.-based corporate educational development contracting a constitutive element of a longstanding U.S. imperial[28] project?

To begin answering this question it is necessary to examine who CAII has worked with in the past, what other organizations CAII is involved with, and how CAII's activities fit into the broader educational dimension of the U.S. national security strategy.

The representation of Kruvant in mass media as principally a symbol of the successes of minority-owned business, as an innocent idealist, maybe even a hippie-holdover that disproves crony capitalism could not be farther from the intersection of multimillion dollar profits and foreign intervention planning work that characterizes Kruvant's enterprise. Kruvant is involved in government policy and Washington, DC business circles having worked on the project in search of a national security strategy. Kruvant sits on the boards of Venture Philanthropy Partners, Calvert Group, and Acacia Federal Savings Bank, and is described by John M. Derrick Jr., chairman and CEO of Pepco, as, "a visible and respected member of the D.C. business community."[29] As such, Kruvant worked to find financial opportunities for large corporations by introducing them to DC small businesses and "From 1996 until 2000, Kruvant served as an emergency schools trustee after the D.C. financial control board stripped the elected board of its powers."[30] These domestic activities of seizing an educational system and representing business interests as democratic governance and philanthropy share a marked resemblance to CAII's international work.

Creative Associates International, Inc. in Iraq, 2003–4

In what follows I discuss the news coverage of corruption and efficient delivery of educational services as well as claims that CAII is involved in implementing "democratic education" in Iraq. I conclude by suggesting that this coverage has obscured the conservative educational privatization agenda, which is possibly the most radical aspect of educational rebuilding and which straddles accumulation by dispossession and accumulation by reproduction.

In March of 2004, a year after the United States invaded Iraq, the Departments of Defense and State released optimistic reports on the situation in Iraq while the Brookings Institution[31] painted a different picture of reconstruction and security in its "Iraq Index," finding that only 65 percent of local security forces are fully-trained, that only 2 percent of the 8,500 "anticoalition suspects" held in detention were foreign nationals despite the claim by State and Defense that the insurgency is composed mostly of "foreign terrorists," monthly electricity levels were less than prewar levels, and only two-thirds of the population had access to potable water.[32] Reporting on these conflicting versions of the state of Iraq the *Atlantic Monthly* wrote, "Perhaps the brightest spot is education: more than 2,300 schools have been rehabilitated by USAID, millions of new textbooks have been printed and distributed, and teachers' salaries are far higher than under the former regime."[33]

Yet in October of 2004 Mary Ann Zehr reported in *Education Week* a very different version of progress on CAII's rebuilding of Iraq's education system as schools reopened in Iraq since the U.S. invasion in March of 2003.

> Results of a ministry survey of schools released this fall show that more than 7,000 of Iraq's 11,000 primary schools either don't have a sewage system at all or don't have one that is operating properly, and that more than 4,000 primary schools have leaking roofs. The survey also estimates that 32,000 additional classrooms are needed.[34]

Despite the Bush administration's No Child Left Behind emphasis on "accountability" of schools, teachers, and students, USAID was hardly forthcoming in response to requests for information on the performance of Creative Associates after the company's first year of operating in Iraq.

> In a March 25 letter, a USAID official justified the rejection of a Freedom of Information Act request for such documents filed by *Education Week* by writing, "Release of this deliberative-process information to the public could hamper the exchange of honest

and open communications and thus adversely interfere with our agency's contract-monitoring activities."[35]

In fact, the secrecy of the performance on the first contract followed the evasion of bidding protocol. In June 2003 USAID conducted an internal investigation and concluded that only one of the five bidders (only CAII) had been invited to the initial discussions with USAID and that procedures had not been followed in awarding the contract. Senator Joseph Lieberman of Connecticut reviewed the investigation and announced that there was "essentially no competitive bidding at all."[36] Congressional critics of the no-bid contracting say that it, "allows the administration to reward friendly companies, prevents Congress from exercising its authority over spending, and may result in higher costs to taxpayers."[37]

In its Iraq work CAII worked with a number of partners including, for example, Research Triangle Institute Inc., which was also involved in forming educational policy at least in the first year of the contract.[38] As part of its activities CAII conducted a survey to discover student-teacher ratios and also designed and manufactured student kits that included pens, pencils, erasers, and notebooks and featured the USAID logo. (Creative Associates International's website inexplicably features a photograph of these kits not being received by students in Iraq but rather being manufactured by a woman in a Chinese factory.) CAII hired Iraqi companies to make furniture for the schools and gave out grant money to set up PTA style organizations. While CAII arranged for the printing and delivery of textbooks, in the first year USAID paid UNESCO $10 million to print 8.6 million math and science textbooks and money from the United Nations "oil for food" program paid UNICEF to print 44.5 million textbooks for all other subjects. Andrew Natsios of USAID announced at a State Department briefing that a group including the Coalition Provisional Authority, the Ministry of Education, UNESCO, UNICEF, were, "working on redoing the textbooks, which were full of vitriol and Baathist party propaganda."[39] "Redoing the textbooks" appears to have involved little more than selective censorship.

For the second year of the U.S. occupation the World Bank joined educational rebuilding offering $40 million for printing the same textbooks and $59 million for school rebuilding. Money originates with World Bank member countries including the United States. The involvement of the World Bank in educational rebuilding is best understood in relation to its longstanding conditions on lending that aims at private sector development through accelerated privatization and liberalization.[40]

For the second CAII contract the role of the company shifted. In an article titled, "Iraq Gets Approval to Control Destiny of School System" (imagine an article that said "Iraq gets Approval to Control Energy Resources!") *Education Week* quoted the Iraqi Minister of Education, a former World Health Organization official, saying that though USAID would continue to play a role in the education rebuilding, control was shifting from the Coalition Provisional Authority to the Ministry.

Simply in terms of "effective delivery of educational services" the performance of CAII is questionable. Williamson M. Evers, a research fellow of the conservative Hoover Institution who worked with the company gave CAII a poor evaluation. Though praising the company for effectively delivering school materials and furniture to schools, and conducting a needs assessment, Evers stated, "All the other things in the contract that had to do with the longer-term development of the Education Ministry—and what is called capacity building—were not done well," he contended. The work "was poor, sloppy, had a lack of follow-through, and a lack of perseverance and persistence."[41] For the second contract CAII dropped American University, American Islamic Congress, and RTI International while adding new subcontractors.

In April 2004 CAII withdrew most of its international staff from Iraq for security. Nidhal Kadhim was an office manager for a CAII subcontractor, Iraq Foundation. One of six Iraqi professionals hired by CAII to advise the Ministry of Education, Kadhim criticized Creative Associates for hiring too few Iraqis under the first contract. In fact, the USAID contract did not obligate CAII to hire any Iraqis. Referring to the company's teacher training project in the first year, Kadhim stated, "They took the whole project to themselves, and it was them who were doing all the materials and doing the training and preparing the materials. . . . It would have been good to have local staff help with the preparation of the materials."[42]

The inefficient delivery of educational services appears inextricably linked to the use of private for-profit corporations. According to Farshad Rastegar, CEO of Los Angeles-based Relief International that has built and repaired schools in Iraq, the failure to utilize nonprofits and instead to use private companies has wasted U.S. federal government money. While 27 cents of every dollar spent on the rebuilding generally has gone for intended projects, according to the Center for Strategic and International Studies, 30 percent is paying for security. Rastegar's group claims to spend only 1 percent on security, "We're not out there in big cars that say, 'I'm an expat, come and attack me.' We're not mixing with the military side of the operation. We're not identified with that."[43]

If the 2003–4 CAII contracts totaled $190 million then the odds are that more than $60 million went not for education at all but for security—most of which was conducted for profit by private security corporations like Blackwater and Dyncorp. And based on the trend of reconstruction spending generally, less than what is spent on security will directly go to education. Is the expensive for-profit educational rebuilding worth the extra money?

> The contract says the purpose of education reconstruction last school year was "to normalize basic education in Iraq following a conflict," but the new contract "focuses on quality and access." To provide that "quality," the contract says, schools will incorporate "democratic practices in the classroom" and develop students' learning and critical-thinking skills.[44]

Much popular press writing on educational rebuilding in Iraq suggests that the emphasis on "democratic education" comes as basic needs of Iraq's children have yet to be met. As tens of millions of dollars are wasted on security to keep private U.S. companies controlling the rebuilding, the status of the youngest and most vulnerable Iraqis is perilous.

In the Summer of 2004 with more than 40 percent of Iraqis below the age of 14, UNICEF found that infant mortality rates had doubled since 1989 just before the first U.S. invasion and the decade of U.S.-led sanctions. "The mortality rate for children under age 5 is two and a half times its 1989 level...children suffer an average of nearly 15 episodes of diarrhea per year, up from 3.8 in 1990, and typhoid cases have spiked from 2,240 to 27,000 in the same period."[45] At the beginning of the 2004–5 school year 5.7 million Iraqi children were expected to attend school yet a national survey showed 7,000 of the 11,000 primary schools did not have a functional sewage system and that 4,000 primary schools have leaking roofs—conditions hardly conducive to children's health.

The point here is not that democratic educational ideals are at odds with these health conditions and that one must choose one or the other. Any democracy requires the health of citizens expected to govern themselves. Rather, the point is that proclamations about spreading democracy are hard to believe as the same people behind the alleged democracy promotion programs have been behind the two invasions, the aerial bombardment, and the devastating decade of sanctions that are estimated to have killed as many as one million Iraqi children directly or indirectly. What is more, the declared intent of democratic education can only make sense in relation to the conditions for democratic governance in Iraq more broadly. The Abu Graib prison situation, attempts to control the outcome of elections in ways favorable to U.S. interests, and the essential theft of Iraqi national wealth for the enrichment of multinational oil companies and for the strategic aims of the United States, seem at odds with an honest effort at democracy promotion. Why the United States would opt for a private company under USAID such as CAII to execute education rebuilding despite the massive monetary waste of using a private contractor, despite the desperate dying children, and despite the massive oil wealth beneath the schools—this can only be understood as being about the retention of U.S. control over the outcome of the education system, an agenda engineered for shifting power and profit to the private sector and retaining U.S. control over Iraqi civil society to implement such an agenda rather than about the will or welfare of the Iraqi people. What is more, if the aim of democratic education is the development of a more democratic society then how should one understand democratic education projects as they are being enacted while democracy is being subverted in a number of other ways—that is, political, economic, and cultural control are being manipulated by the United States rather than being controlled by the Iraqi people.

Most pertinent here, the declaration of building "democratic education" can only be understood in relation to other declared aims of CAII's second contract. These other aims involve privatization. In April of 2004 Zehr reported that CAII had been awarded a second contract from USAID with a different mission than that of the first contract. The first contract called for CAII to distribute furniture and materials to schools, to train about 33,000 teachers in "student-centered" educational methods, administer a survey to evaluate the needs of secondary schools, create accelerated learning programs for 600 students, distribute grants for repairs to schools, and establish an information management system for the Ministry of Education.[46]

The second contract appears to set the stage for privatization of the Iraqi education system through "strengthening a decentralized education structure."[47] Such "decentralization" would advance the goal of nurturing "public-private partnerships," something that USAID makes explicit on its website.[48] The model for this appears to be the

growing U.S. charter school movement that the federal government
of the United States has supported with billions of dollars. Charters
are the spearhead of public school privatization as more than three
quarters of new charters opened by for-profits are charter schools.[49]
Domestically in the United States, one of the three major thrusts of
"No Child Left Behind" is charter school promotion. The charter
movement following the ideals of neoliberal economic policy emphasizes decentralization, deregulation, experimentation, involvement of
the private for-profit sector, the undermining of teachers unions and
local democratically elected school councils, and the handing of management over to business groups. Though the U.S. media has largely
failed to pick up on the attempts to remake education on the current
conservative educational reforms, the Assyrian International News
Agency reports that a new crop of private for-profit schools are being
opened in Iraq. Saddam Hussein had nationalized education in 1973
and Iraq was regarded as having one of the best education systems in
the Middle East with full gender inclusion, free to all, fostering
80 percent literacy, and a secular curriculum that did not require non-
Muslims to partake in religious instruction. The wars with Iran that
the United States fueled from both sides, the Gulf War, and the decade
of U.S.-led UN sanctions destroyed the educational system with the
U.S. invasion being the final straw.

> Certifying private schools is a way to add classrooms without tap-
> ping public coffers, [Interim Minister of Education] Allaq said.
> After years of surviving on subsidies, "The citizen is realizing
> that not everything can be provided by the government," he said.
> Private schools also received a boost because some of the American
> advisers sent to work with Iraq's transitional government had ties
> to the U.S. charter school movement and supported more local
> control of Iraqi schools.[50]

Allaq's parroting of the neoliberal U.S. justification for educational privatization is hard to fathom when one considers the amount of money
being allocated for military, policing, and other repressive measures
that are principally necessary because of the continuing U.S. military
presence. Though the noncompetitive bidding between the federal
government and a for-profit corporation does raise important questions
about corruption and how the public sector is being used to enrich a
tiny elite in the private sector, this issue tends to eclipse a more fundamental one. Namely, that the broader issue appears to be the role of

military destruction and reconstruction as a form of neocolonialism—a way to justify the privatization agenda.

"We used to have vulgar colonialism," says Shalmali Guttal, a Bangalore-based researcher with Focus on the Global South. "Now we have sophisticated colonialism, and they call it 'reconstruction.'"...If anything, the stories of corruption and incompetence [in rebuilding] serve to mask this deeper scandal: the rise of a predatory form of disaster capitalism that uses the desperation and fear created by catastrophe to engage in radical social and economic engineering.[51]

It is just this radical social and economic engineering that appears to be the real story that educators committed to democratic ideals need to pay attention to as media coverage of this issue appears nearly nonexistent. As in the domestic debates over corruption and efficient delivery of for-profit educational services (such as the Edison Schools), the preponderance of press coverage focusing on fair business practices conceals the broader implications that privatization of public education has for a democratic society. Privatization of public education shifts control over public institutions to private hands thereby undermining the role public education plays as a democratic public sphere, as a space for public deliberation over values, meanings, and matters of public import. Privatization of public education also redefines the very meaning of the public in private terms by treating a service that matters for the whole society as a consumable commodity that matters only for the individual consuming unit. Privatization also redefines public citizens as private consumers.

As long as the United States oversees and influences the new Iraqi Ministry of Education and USAID and the World Bank continue to rebuild, U.S. models of control can be expected to be followed. Mechanisms of formal democracy promoted by "democracy promotion" projects are more conducive to corporate globalization than would be projects designed to foster democratic control over the means of production, democratic decision making over consumption, or democratic control over meaning-making technologies (schools, media, religion), which would shift power away from for-profit institutions and toward public control. Educators and cultural workers committed to such genuinely democratic shifts in control ought to not only oppose the U.S. occupation but understand that the authoritarian privatization trends of the "new imperialism" and "disaster capitalism" can be identified operating not just in the "nonintegrating gap" but also in those power centers in the nations waging imperial war.

Notes

Excerpts above from Kenneth Saltman, "Creative Associates International: Corporate Schooling and 'Democracy Promotion' in Iraq," *Review of Education, Pedagogy and Cultural Studies* 28 (2006): 25–65 reprinted by permission of Taylor and Francis.

1. See Irvin Stelzer, "Introduction," in *The Neocon Reader*, ed. Irvin Stelzer (New York: Grove Press, 2004), 3–28. The vision of remaking the world on the singular American model is clearly stated in the opening of the national security strategy of the United States available at (www.whitehouse.gov). The publication of *The Neocon Reader* was launched with an event covered by C-Span television and included lectures by William Kristol and Irvin Selzer. The early plans for the reformulation of U.S. national security strategy can be found on the neoconservative Project for a New American Century website.

2. Despite repeated assertions by the Bush administration that the invasion of Iraq in March of 2003 was part of the "War on Terrorism" no credible links have been found between the Islamist political movement "Al Qaeda" and the Iraq of Saddam Hussein. Osama Bin Laden and Saddam Hussein were bitter enemies with incompatible ideological convictions. The admission by the Bush administration that no weapons of mass destruction could be found in Iraq revealed as false the original justification for the war, which was the immediate security of the United States. This makes the war an illegal act within international law. In the build up to war and since the invasion the justifications for war by the administration and the news media have been interchangeably security, democratic nation-building, moral imperative, and paternal revenge. In the context of education these interchangeable justifications have been well-documented on Megan Boler's website Critical Media Literacy and War available at http://www.ncr.vt.edu/mediaproject/home.htm (accessed December 10, 2008).

3. Jackie Spinner, "Questions Raised about Iraq Contract," *The Washington Post*, June 13, 2003, Financial, E2. David Morris, "Criticism Grows of No-Bid Work for Iraq Reconstruction," *Congress Daily*, April 16, 2003, 3.

4. This is detailed in Pratap Chatterjee, *Iraq, Inc.: A Profitable Occupation* (New York: Seven Stories Press, 2004). See also Christian Parenti, "Fables of the Reconstruction," *The Nation*, August 30/September 6, 2004, 16–19.

5. "War Profiteers" is how corporate watchdog group Corp Watch described the rebuilding contractors.

6. Christian Parenti describes the rebuilding generally this way.

7. "Windfalls of War" is the label provided by the Center for Public Integrity.

8. Stelzer, "Introduction."

9. A more extensive version of this article including sections on Haiti and Nicaragua appears as a chapter in Kenneth J. Saltman, *Capitalizing on Disaster: Taking and Breaking Public Schools* (Boulder, CO: Paradigm Publishers, 2007) and as an article in Kenneth J. Saltman, "Creative Associates International: Corporate Education and 'Democracy Promotion' in Iraq," *The Review of Education, Pedagogy, and Cultural Studies* 28 (2006): 25–65.

10. William I. Robinson has written extensively and importantly on the U.S. foreign policy shift away from support for and promotion of authoritarian regimes and toward promotion of what he terms *polyarchy*, forms of democracy that ratify elite rule through formal democratic processes while averting popular rule and control and assuring market economies more effectively than authoritarianism. See for example, William I. Robinson, *Promoting Polyarchy* (Cambridge: Cambridge University Press, 1996) and William I. Robinson,

"Globalization, The World System, and 'Democracy Promotion' in US Foreign Policy," *Theory and Society* 25 (1996): 615–665.

11. See for example, Pratap Chatterjee, *Iraq, Inc.: A Profitable Occupation* (New York: Seven Stories Press, 2004).

12. I have taken up the relationships between recent educational reform and privatization in *The Edison Schools: Corporate Schooling and the Assault on Public Education* (New York: Routledge, 2005).

13. As David Harvey and Giovanni Arrighi note if the prominent figures in the Clinton administration were the finance people such as Rubin and Summers then it is the military people in the Bush White House such as Rumsfeld, Cheney, and the neoconservative staff of hawks like Perle Wolfowitz, etc. See David Harvey, *The New Imperialism* (Oxford: Oxford University Press, 2003) and Giovanni Arrighi, "Hegemony Unravelling," *New Left Review* 32 (2005): 23–80.

14. Robinson, "Globalization," 619.

15. Leslie Sklair, "The Culture-Ideology of Consumerism in the Third World," chap. 5 in *Sociology of the Global System* (Baltimore, MD: Johns Hopkins University Press, 1991), 147–190.

16. Henry Giroux has importantly and correctly criticized the theoretical shortcomings of Althusser's project in the context of education. See especially *Theory and Resistance in Education* and *Education Still Under Siege* for these crucial criticisms. Despite Althusser's shortcomings his insights about state power offer distinct tools for the present historical juncture.

17. This is not to discount the systematic ways the U.S. waged war on the Third World from the end of World War II to the present particularly in Central America, but from the end of the Vietnam War and the coinciding rise of neoliberalism.

18. See Kenneth J. Saltman and David Gabbard, *Education as Enforcement: The Militarization and Corporatization of Schools* (New York: Routledge, 2003). From public schools made into military academies to JROTC to a rising punitive culture of discipline this phenomenon is only increasing particularly as the U.S. military becomes desperate to fill its ranks.

19. See Kenneth J. Saltman, *Collateral Damage: Corporatizing Public Schools—A Threat to Democracy* (Lanham, MD: Rowman & Littlefield Publishers, 2000).

20. The Center for Public Integrity, "Windfalls of War: Creative Associates International Inc." available at http://publicintegrity.org/wow/bio.aspx?act=pro&ddlC=11 (accessed December 10, 2008).

21. Ibid.; Jackie Spinner, "Iraq: Operation Iraqi Education," *Washington Post*, April 21, 2003.

22. Center for Public Integrity, "Windfalls of War: Creative Associates International Inc."

23. This is how the Center for Public Integrity describes her teaching past.

24. Center for Public Integrity, "Windfalls of War: Creative Associates International Inc."

25. Spinner, "Operation Iraqi Education."

26. Vice President Cheney headed Haliburton and continued to benefit economically from the company. Haliburton's subsidiaries include Bechtel. Condaleeza Rice worked for Chevron, which named an oil tanker after her. Iraq rebuilding administration shifted to the National Security Agency under Rice and Chevron received early large oil contracts following the invasion.

27. Staff, "The Spoils of War: Cleaning Up," *The Economist*, April 3, 2003.

28. For my use of the term *imperialism* I draw on the work of David Harvey, *The New Imperialism*, Douglas Stokes' essays, Ellen Meiskins Wood, *Capitalist Imperialism*, Michael Parenti's *Against Empire* as well as Hannah Arendt's *The Origins of Totalitarianism*. These more or less nuanced versions of empire emphasize that while the nation-state may be weakened by the deregulatory rules of the post-fordist economy, they emphasize the tensions between the extra-national interests of a transnational capitalist class and the wielding of national power for economic advantage by military action against other nations. Parenti writes, "By

'imperialism' I mean the process whereby the dominant poitico-economic interests of one nation expropriate for their own enrichment the land, labor, raw materials, and markets of another people" (1). The term imperialism, long derided in mass media as little more than a loony left marker of conspiracy theory and rejected as illegitimate scholarship in academia has only recently again begun to be taken seriously across the political spectrum as a traditional conservative isolationism joins progressive and radical left criticism of the neoconservative plans for U.S. global military control.

29. Spinner, "Operation Iraqi Education."
30. Ibid.
31. Though the Brookings Institution is regarded as a politically "centrist" think-tank it is clearly in favor of conservative educational privatization plans and counts as fellows a number of outspoken advocates of for-profit public schooling such as John Chubb the Chief Education Officer of The Edison Schools. This is discussed at length in Kenneth J. Saltman, *The Edison Schools: Corporate Schooling and the Assault on Public Education* (New York: Routledge, 2005).
32. Primary Sources, "Foreign Affairs: Iraq by the Numbers," *Atlantic Monthly* 294, no.1 (2004): 60.
33. Ibid.
34. Mary Ann Zehr, "Schools Open in Iraq, After Two-Week Delay," *Education Week* 24, no.7 (2004):6–7.
35. Mary Ann Zehr, "Iraq Gets Approval to Control Destiny of School System," *Education Week* 23, no. 31 (2004): 1.
36. Ibid.
37. David Morris, "Criticism Grows of No-Bid Work for Iraq Reconstruction," 3.
38. CAII subcontracted Iraq rebuilding work with three of the four companies that were invited to bid but did not. These included Research Triangle Institute and DevTech Systems Inc. Research Triangle in turn subcontracts to CAII. CAII also subcontracts to American University, American Manufacturers Export Group, Booz Allen Hamilton, and Camp Dresser & McKee International and two non-profits led by Iraqi expatriates, American Islamic Congress and the Iraqi Foundation.
39. Center for Public Integrity, "Windfalls of War: Creative Associates International Inc," 4.
40. See Steven J. Klees, "The Implications of the World Bank's Private Sector (PSD) Strategy for Education: Increasing Inequality and Inefficiency," Citizens' Network on Essential Services http://servicesforall.org/html/tools/Klees_PSD_Paper_1–15-02.shtml.
41. Zehr, "Iraq Gets Approval to Control Destiny of School System," 3.
42. Mary Ann Zehr, "Creative Associates Gets New Iraq Contract," *Education Week* 23, no. 42 (2004): 17.
43. Mary Ann Zehr, "Schools Open in Iraq After Two-Week Delay," 6.
44. Ibid., 2.
45. Valerie J. Brown, "Reconstructing the Environment in Iraq," *Environmental Health Perspectives* 112, no.8 (2004): A464.
46. Zehr, "Iraq Gets Approval to Control Destiny of School System," 1.
47. Zehr, "Creative Associates Gets New Iraq Contract," 17.
48. USAID's website has a number of links to a number of programs highlighting its emphasis on privatizing public sector provision. There are explicit programs for development of private sector involvement in education while neoliberal ideology informing the perspective of USAID celebrates liberalization and privatization of service sector.
49. Alex Molnar, Glen Wilson, and Daniel Allen, "Profiles of For-Profit Educational Management Companies: Fifth Annual Report," Commercialism in Education Research Unit at Arizona State University, http://www.eric.ed.gov/ERICWebPortal/contentdelivery/servlet/ERICServlet?accno=ED480738.

50. Robin Fields, "Iraq Ministry of Education Withholds Approval for Private Assyrian School," *The Los Angeles Times*, reprinted at www.aina.org (accessed December 10, 2008).
51. Naomi Klein, "The Rise of Disaster Capitalism," *The Nation*, May 2, 2005, 9.

Bibliography

Arendt, Hannah. *The Origins of Totalitarianism.* New York: Harcourt, Brace & World, 1966.

Arnowitz, Stanley and Henry A. Giroux. *Education Still Under Siege.* 2nd ed. Westport, CT: Bergin and Garvey, 2001.

Arrighi, Giovanni. "Hegemony Unravelling." *New Left Review* 32 (2005): 23–80.

Boler, Megan. "Critical Media Literacy and War." Department of Teaching and Learning, Virginia Tech. http://ncr.vt.edu/mediaproject/ (accessed December 10, 2008).

Chatterjee, Pratap. *Iraq, Inc.: A Profitable Occupation.* New York: Seven Stories Press, 2004.

Giroux, Henry A. *Theory and Resistance in Education: A Pedagogy for the Opposition.* Westport, CT: Bergin and Garvey, 2001.

Harvey, David. *The New Imperialism.* Oxford: Oxford University Press, 2003.

Klees, Steven J. "The Implications of the World Bank's Private Sector (PSD) Strategy for Education: Increasing Inequality and Inefficiency." Citizens' Network on Essential Services. http://servicesforall.org/html/tools/Klees_PSD_Paper_1–15-02.shtml (accessed December 10, 2008).

Molnar, Alex, Glen Wilson, and Daniel Allen. "Profiles of For-Profit Educational Management Companies: Fifth Annual Report." Commercialism in Education Research Unit at Arizona State University. http://www.eric.ed.gov/ERICWebPortal/contentdelivery/servlet/ERICServlet?accno=ED480738 (accessed December 10, 2008).

Parenti, Christian. "Fables of the Reconstruction." *The Nation*, August 30/September 6, 2004.

Parenti, Michael. *Against Empire.* San Francisco: City Lights Books, 1995.

Robinson, William I. *Promoting Polyarchy: Globalization, U.S. Intervention, and Hegemony.* Cambridge: Cambridge University Press, 1996.

———."Globalization, The World System, and 'Democracy Promotion' in US Foreign Policy." *Theory and Society* 25 (1996): 615–665.

Saltman, Kenneth J. *Capitalizing on Disaster: Taking and Breaking Public Schools.* Boulder, CO: Paradigm Publishers, 2007.

———. *Collateral Damage: Corporatizing Public Schools—A Threat to Democracy.* Lanham, MD: Rowman and Littlefield Publishers, 2000.

———. "Creative Associates International: Corporate Education and 'Democracy Promotion' in Iraq." *The Review of Education, Pedagogy, and Cultural Studies* 28 (2006): 25–65.

———. *The Edison Schools: Corporate Schooling and the Assault on Public Education.* New York: Routledge, 2005.

Saltman, Kenneth J. and David Gabbard. *Education as Enforcement: The Militarization and Corporatization of Schools.* New York: Routledge, 2003.

Sklair, Leslie. "The Culture-ideology of Consumerism in the Third World." Chap. 5 in *Sociology of the Global System.* Baltimore, MD: Johns Hopkins University Press, 1991.

Stelzer, Irvin. "Introduction," in *The Neocon Reader,* ed. Irvin Stelzer, 3–28. New York: Grove Press, 2004.

Williams, Brooke. "Windfalls of War: Creative Associates International Inc." The Center for Public Integrity. http://publicintegrity.org/wow/bio.aspx?act=pro&ddlC=11 (accessed December 10, 2008).

Contributors

Dana Burde is an assistant professor in the Steinhardt School of Education at New York University and formerly was a research scholar at the Saltzman Institute of War and Peace Studies in the School of International and Public Affairs at Columbia University. Her research and teaching focus on education in emergencies, nongovernment organizations, and social movements, and on education as a tool for social reconstruction in post-conflict regions. Her current research focuses on the link between education, protection, and life chances for adolescents living in post-conflict societies and is supported by grants from the Spencer Foundation, the U.S. Institute of Peace, and the Weikart Foundation. Recent publications include: *Education in Crisis Situations: Mapping the Field* (December 2005), Washington DC: Basic Education Support Project/USAID; *Save the Children's Afghan Refugee Education Program in Balochistan, Pakistan 1995–2005.* (2006), Westport, CT: Save the Children. Beyond the university, her work as an educational consultant includes assessment and evaluation of post-conflict programs in the Balkans; civil society building in the Caucasus; refugee education in Pakistan; and research on parent and community participation in community schools in Central America and Mali. Burde received her PhD in Comparative and International Education from Columbia University; Ed.M. in Administration, Planning, and Social Policy/International Education from Harvard University; and BA from Oberlin College.

Charles Dorn is an assistant professor of Education and chair of the Education Department at Bowdoin College in Brunswick, Maine. He received his MA from Stanford University in 1994 and his PhD from the University of California, Berkeley, in 2003. Dorn's research into the history of education investigates the civic functions adopted by and ascribed to educational institutions in the United States, including centers of early childhood education, public elementary and secondary schools, and colleges and universities.

Benjamin Justice is an associate professor of Education and (by courtesy) History and codirector of the Rutgers Social Studies Education Program. He was 2005–6 National Academy of Education/Spencer Post Doctoral Fellow, researching the history of Americans' use of education in nation building. He is the author of *The War That Wasn't: Religious Conflict and Controversy in the Common Schools of New York State, 1865–1900* (SUNY 2005), as well as various book chapters in edited volumes. His work has appeared in the *History of Education Quarterly, Teachers College Record, Social Education, Philosophy of Education,* and *New York History.*

Thomas Koinzer, born in 1967 in Arnstadt/Thuringia, studied History, Sociology and Political Sciences in Halle, Bamberg, Berlin, at Duke University and Essex University, PhD in 2001 Humboldt University Berlin, senior lecturer at the Institute of Educational Sciences, Humboldt University Berlin.

M. Ayaz Naseem is an assistant professor at Concordia University in Montreal, Canada. He is the author of a book on Pakistani-Soviet relations (Progressive Publishers 1988). His scholarship on Pakistani education has appeared in edited volumes and in journals such as the *International Journal of Inclusive Education.*

Kentaro Ohkura is an associate professor of Comparative Education and Social Theory at Tamagawa University, Tokyo, Japan. He is currently conducting research that reexamines the U.S. and Japanese policies of education in terms of social inclusion/national integration.

Laura B. Perry is lecturer of education policy at Murdoch University in Perth, Australia. She is currently developing a theoretical model for conceptualizing democratic education, and analyzing cross-national differences in such conceptions. Her work has appeared in journals such as *Compare.* She received her PhD in education policy and comparative education from Loyola University Chicago.

Brian Puaca is assistant professor of Christopher Newport University. He received his PhD from the University of North Carolina—Chapel Hill. His dissertation, titled "Learning Democracy: Education Reform in Postwar West Germany, 1945–1965" was awarded the History of Education Society Award for best dissertation in 2006.

Kenneth J. Saltman is associate professor at DePaul University in Chicago. He has written extensively on public school privatization and corporate involvement in schooling. His books include *Collateral Damage: Corporatizing Public Schools—A Threat to Democracy* (2000), *Strange Love, Or How We Learn to Stop Worrying and Love the Market*, with Robin Truth Goodman (2002), *Education as Enforcement: The Militarization and Corporatization of Schools*, with David Gabbard (2003) and *The Edison Schools: Corporate Schooling and the Assault on Public Education* (2005).

Masako Shibata is assistant professor in the Graduate School of Humanities and Social Sciences, University of Tsukuba, Japan, and editor of *Research in Comparative and International Education*. Her publications on the theme of educational transfer include *Japan and Germany under the U.S. Occupation* (Lexington Books, 2005).

Noah W. Sobe is assistant professor of Cultural and Educational Policy Studies in the School of Education at Loyola University Chicago, where he is also associate director of Loyola's Center for Comparative Education. His research looks historically at the international circulation of curricula and pedagogical theories. He received his PhD from the University of Wisconsin–Madison and his MA from Teachers College, Columbia University. His publications include articles in *Paedagogica Historica, Educational Theory, European Education, Harvard Education Review, Current Issues in Comparative Education (CICE)* and the book, *Provincializing the Worldly Citizen: Yugoslav Student and Teacher Travel and Slavic Cosmopolitanism in the Interwar Era* (Peter Lang, 2008).

Jason M. Yaremko is an assistant professor with the Department of History, University of Winnipeg, and History Program Coordinator with the Faculty of Education's Five-Year Bachelor of Education Program, an Access program run by the University of Winnipeg. He teaches the History of the Americas, and his research focuses on intercultural relations in Cuba and other areas of Latin America, as well as on the role of religion, the state, and education in the acculturation or transculturation process. His publications include *U.S. Protestant Missions in Cuba: From Independence to Castro* (University of Florida, 2000) and a number of articles.

INDEX